WARBIRDS OVER WANAKA

THE STORY OF NEW ZEALAND'S PREMIER AIRSHOW

WARBIRDS OVER WANAKA

THE STORY OF NEW ZEALAND'S PREMIER AIRSHOW

GERARD S. MORRIS

Foreword by Tim Wallis

REED

Published by Reed Books, a division of Reed Publishing (NZ) Ltd,
39 Rawene Road, Birkenhead, Auckland.
Associated companies, branches and representatives
throughout the world.

ISBN 0 7900 0327 9

© Gerard Morris 1994
First published 1994

Cover and text design by Chris Lipscombe
Printed in Hong Kong

Front cover photograph The Dogfight: Spitfire vs Messerschmitt, 1992.
Back cover Mirror formation, RNZAF Red Checkers display.
Opposite title page Fury ZK-JHR displays its classic lines, in the RAF colours of K2059.
Contents 'Plonky', the insignia on the Grumman Avenger ZK-TBM.
Foreword Tim Wallis displays his skills in ZK-HOT.

This book is dedicated to Judy,
my parents Derek and Margaret Morris,
my brothers and sisters and their families.

Gerard Morris was born in Invercargill, New Zealand, and
has had a keen interest in aviation for as long as he can
remember. He received his secondary school education at
Southland Boys' High School, Invercargill. On leaving school
he joined the then Government Audit Department and was
based in the Invercargill office for over ten years. He
graduated from the University of Otago with a Bachelor of
Commerce degree, majoring in accounting. In 1993 he
studied at the Christchurch College of Education toward a
Diploma in Teaching.

Contents

Acknowledgements

Very special thanks must go to Tim Wallis, Managing Director of the Alpine Fighter Collection and Chairman, Warbirds Over Wanaka. Without his support, this book would not have been possible. Thanks also to Ray Mulqueen, Chief Engineer of the AFC, and Ian Brodie, Manager of the New Zealand Fighter Pilots Museum, who have shared my enthusiasm and been only too willing to give their time and valuable information.

The special contributions from the following people were very much appreciated: Stephen Grey of The Fighter Collection, Duxford, England; Mark Hanna, of the Old Flying Machine Company, Duxford, England; Air Vice-Marshal John Hosie OBE, RNZAF, Chief of Air Staff, Defence Headquarters, Wellington; Trevor Bland, President of the New Zealand Warbirds Association, Auckland.

To John Lamont, Chief Pilot for the Alpine Fighter Collection, Warren Russell, a member of the Aviation Historical Society of New Zealand, and Alan Falconer, an aviation enthusiast, thanks for proofreading. Kevin Bennett, of the Aviation Historical Society of New Zealand, confirmed the aircraft details provided in the appendices.

I would like to thank especially the Thorburn family, Graeme and Robyn and their sons Marcus and Stefan (Dunedin). They let me bounce ideas off them and shared my enthusiasm throughout the two years it took to put this book together.

Thanks must also go to: Gavin Johnston (Airshow Organising Committee), Lloyd Dunn (former Chairman, Wanaka Airport Authority), Stan Kane (former Chairman Vincent County Council), Simon Spencer-Bower (Civil Air Display Planning Committee), Brian Farrell (Civil Aviation Authority of New Zealand, Wellington), Squadron Leader J.D. Kirtlan (RNZAF Base, Wigram), Squadron Leader Paul Harrison (Headquarters, New Zealand Defence Force, Wellington), Squadron Leader Ian McClelland (RNZAF Base, Wigram), Flight Lieutenant A.D. Percy (RNZAF Base, Wigram), Flight Sergeant Michael Provost (RNZAF Base, Wigram), Howard Monk (Secretary, New

Zealand Warbirds Association, Ardmore), Ross Macpherson (*New Zealand Wings*), Colin Smith (Croydon Aircraft Company, Mandeville), John Swan (Typesetting and Design Ltd, Dunedin), David Duxbury, Errol Martyn and David Bates (Aviation Historical Society of New Zealand), Charlie Kenny (New Zealand Amateur Aircraft Constructors Association), Brett Kennedy (Dunedin), Alan Gibb (Invercargill), Bryan Beck (Mosgiel), Hank Sproull (Queenstown), Doug Maxwell (Alexandra), Stephen Low (Mosgiel), Doug Hopkins (Dunedin), Aidan Bourne (Dunedin), Peter Brown (Dunedin), Felicia Yii (Wellington), Paul Squire (Dunedin), Paul Sortehaug (Dunedin), Laurie Thomas (Timaru), Michelle Longland (Christchurch), Michael Holmes (Christchurch Small Business Bureau Ltd), Shafid Khan (Khan Software, Dunedin), my friends at the University of Otago and Kings High School Hostel, Dunedin, and Sonoda Christchurch Campus, Christchurch. My 1993 Professional Studies Tutorial Group and staff at the Christchurch College of Education.

And to all those unnamed, please accept these words as my acknowledgement of your assistance.

Abbreviations

Bu. No.	Bureau Number
c/n	constructor's number
DFC	Distinguished Flying Cross
RAAF	Royal Australian Air Force
RAF	Royal Air Force
RCAF	Royal Canadian Air Force
RN	Royal Navy
RNZAF	Royal New Zealand Air Force
USAAF	United States Army Air Force
USN	United States Navy

Foreword

From small beginnings Warbirds Over Wanaka has grown to become an international event, supported by a large number of people who travel great distances to take part. Now the success of this airshow has been placed on permanent record through the enthusiasm and initiative of Gerard Morris.

There are many things that make Warbirds Over Wanaka unique, not least the wonderful setting of Wanaka Airport, with its backdrop of glacial terraces and mountains, and the winding Clutha River, providing a dramatic setting for the warbirds to show their mettle once again.

Even more vital, though, is the dedication of the pilots, the mechanics, and all those without whom there would be no airshow. Warbirds Over Wanaka is very much a community event. The local Wanaka people and service clubs give it their enthusiastic support, as do the sponsors, the aircraft owners and the pilots, not to forget the New Zealand Warbirds Association, the RNZAF, the Air Training Corps, Police and Fire Brigades, and the many exhibitors who contribute to the success of the event. The spectators, too, are an essential part of the occasion, for without their support there would be no show.

It is for all these people that Gerard has written this book, and I congratulate him on his achievement. The challenge for the Warbirds team is to make each show more exciting and dynamic than the last. Gerard's challenge has been to record the essence of each show, and the results are here for all to see and enjoy.

Tim Wallis
Chairman
Warbirds Over Wanaka

**New Zealand Warbirds
Association (Inc)**
The badge was designed by Ernie
Thompson in 1978. It depicts an eagle
about to land and is symbolic of the
aircraft operated by New Zealand
Warbirds Association members
representing the two World Wars.

Contribution from Trevor Bland

Since its inception in 1978, the New Zealand Warbirds
Association has established itself as a major contributor
to the arena of restoration and display of vintage and
classic aircraft. Through syndication and the enthusiasm of its
members, the Association continues to grow at a steady rate,
with an average of three to four new aircraft each year.

We as a team are privileged to be part of the Warbirds Over
Wanaka airshow, which has now gained recognition on the
International Airshow circuit. The Organising Committee have
excelled in providing the aviation enthusiast and the general
public with an airshow spectacular without equal and this
book is a fitting tribute to their efforts.

The hospitality and the flying conditions experienced at
Wanaka must be the envy of aviation organisations all over the
world, and provide the vehicle for aviation buffs to mix and
mingle over an exciting three-day period.

Thank you to Tim and Prue Wallis and their 'TEAM' and
thank you to Gerard for giving Warbirds Over Wanaka the
recognition it deserves.

Trevor Bland
President
NZ Warbirds Association (Inc)

Contribution from
Air Vice-Marshal J.S. Hosie

When the RNZAF sold the Harvards in 1977, who could have predicted that they would have formed the basis of an ever-growing 'Warbirds' aircraft movement in New Zealand? Since the sale of the Harvards, the RNZAF has maintained close links with the groups and individuals that have sought to acquire and fly ex-military aircraft that have served the RNZAF or with overseas military forces.

The biennial Warbirds Over Wanaka series of airshows has become a focal point for the display of New Zealand's warbird collection to enthusiasts from here and abroad. The RNZAF has contributed to these airshows, showing the continuing links from past military aircraft to the current range of aircraft serving the Air Force.

The Air Force looks forward to continuing its association with the splendid Warbirds Over Wanaka airshows.

J.S. Hosie OBE
Air Vice-Marshal, RNZAF

Introduction

Wanaka is nestled among the hills of the central South Island of New Zealand, on the southern shores of a beautiful lake. It was first settled by Europeans in 1867, and named Pembroke after a British Colonial Secretary of the 1850s. The town became a centre for the area's timber milling, gold mining, sheep farming and rabbit farming.

In September 1940 the residents of Pembroke's 100 or so households decided to change their town's name to that of the lake, Wanaka. They wanted their town to become a centre of tourism, and so a new image was needed. Typical of Wanaka's new-found enthusiasm was the Wanaka Hotel's advertisements in Dunedin, urging prospective visitors to stay 'a thousand feet above worry level'.

Today, Wanaka attracts over 50,000 tourists annually (not including airshow spectators), who come to experience a wide range of activities, including fishing, kayaking, rafting, snow skiing and water skiing, and to visit attractions such as the maze, the Alpine Fighter Collection and the New Zealand Fighter Pilots Museum.

For three days every two years Wanaka's resident community of 1,700 submits its uncomplicated and relaxed lifestyle to a rush of thousands of people, young and old. The Easter rush is not for the gold of yesteryear but for a different metal, a metal crafted into the shape of flying machines. The Warbirds Over Wanaka Airshow creates an atmosphere of excitement tinged with danger. It recaptures the sense of adventure which led New Zealanders to be among the first in the world to fly.

People travel from overseas and from all parts of New Zealand to attend this now internationally acclaimed event. The success of the airshow is due largely to the efforts of one man, Wanaka-based businessman Tim Wallis. Tim first took to the skies in 1965, flying helicopters on deer recovery operations, and to date has logged over 12,000 hours. In 1968 a helicopter accident left him in Christchurch's Burwood Hospital spinal unit with a broken back. He was told he would not walk again,

Above Tim Wallis.
Previous page Lake Wanaka, with the distinctive hills in the background.

but through sheer grit and determination he continued to run his three companies from his bed. That is typical of Tim's fighting spirit.

And he did walk again. He also never forgot his stay in hospital, as he showed in 1990 when, with only two hours' notice to the hospital, he put on a display in his beloved Spitfire. Patients and staff loved it; unfortunately, it also attracted the attention of the Civil Aviation Authority, who prosecuted Tim. The judge threw the case out of court.

In 1984 Tim Wallis purchased a P-51D Mustang from the United States (sold 1986) and the Alpine Fighter Collection (AFC) was born. The flying collection has grown over the years to include a de Havilland DH82A Tiger Moth, a de Havilland DHC-1 Chipmunk, a Vickers Supermarine Spitfire XVIe, a Curtiss P-40K Kittyhawk, a Chance Vought F4U-1 Corsair, a Grumman TBM-3E Avenger, an A6M2 Zeke (modified Harvard airframe), a Yakovlev Yak 3 near completion in Russia, a Focke-Wulf 190A-8 under construction in Florida, a Polikarpov I-16 under construction in Russia, a North American P-51D Mustang in joint ownership with Brian Hore, and the possibility of a Messerschmitt 109. Waiting in the hangar for possible restoration is a second P-51D Mustang (NZ2427).

Bring together a staggering assortment of exotic aircraft including a Sea Fury, a Mustang, a Venom, a Vampire, a DC-3, a Dragon Rapide, Harvards, Tiger Moths, plus overseas visitors such as a Messerschmitt, and the stage is set for a great display. When some of New Zealand's and England's best display pilots take the controls you have an airshow second to none in the Southern Hemisphere.

The organising committee of Tim Wallis, George Wallis and Gavin Johnston have worked tirelessly over the years to bring continuing excitement to the enthusiastic crowds that grow by almost 20,000 biennially. I believe the time will come when the airshow crowd will exceed 100,000. This is the sort of encouragement on which the aircraft owners thrive. They like to see their hard work in maintaining the precious relics of early aviation being appreciated.

The support of the New Zealand Warbirds Association, local aircraft operators and other aviation groups around New Zealand plus continued sponsorship by corporations, especially

Air BP, Air New Zealand and P & O New Zealand, must also be acknowledged.

The AFC's first ten years have been the source of much media and public attention; the next ten will be even more so. With Russia opening its boundaries, an exciting array of warbirds are becoming available and with Tim's continued determination some of these are likely to appear in New Zealand skies. The combination of Alpine Fighter Collection aircraft and privately owned warbirds which are being encouraged to make Wanaka their home — such as Gerald Rhodes's DH104 Devon and Canberra and a MiG 15 from Australia — is creating a collection that is fast becoming the Duxford of the Southern Hemisphere.

This book has been written because of the increasing popularity of the Warbirds Over Wanaka airshows, and because I felt a record of these shows was required that was more permanent than newspapers, magazine articles and people's memories. The photographs have come from some of New Zealand's best photographers, using a variety of cameras and film. With the exception of the Dunedin shots of the Messerschmitt assembly and the Mustang/Corsair, all photos were taken at Wanaka Airport. Uncredited photographs were taken by the author. All aircraft that contributed to the flying displays are listed in the appendices, with their registration numbers, display pilots, owners and base at the time of airshow(s).

Alpine Fighter Collection

The insignia of the Alpine Fighter Collection shows a warrior holding a taiaha, set on an RAF roundel. The effigy of the Maori warrior has been adopted from the squadron crest of No. 485 (New Zealand) Squadron, which flew Spitfires during World War II. The squadron's blazon noted: 'The ancient Maori warrior had a reputation second to none as a courageous and gallant fighter who loved to fight for fighting's sake. The gesture of the Maori on the badge is symbolic of a besieged and starving tribe's defiance to a call for surrender, symbolising the squadron's spirit.'

In the more culturally aware 1990s, a more appropriate blazon would be: 'The ancient Maori warrior had a reputation second to none as a courageous and gallant fighter. The gesture of the Maori on the badge is the stance of a warrior about to parry the thrust of an opponent with a countermove. The badge is symbolic of the squadron's wartime role.'
(*Thanks to Tom Rangi, Head of Department, Maori Studies, Christchurch College of Education.*)

Wanaka Airport — The Beginning

Today's Wanaka Airport, 7 km from the township, an hour's drive from Queenstown and only three and a half hours from Invercargill, Dunedin and Timaru, can trace its origins back over 60 years of flying in the district.

Aviation appears to have had a strong following among the pioneer farmers, who as early as the 1930s recognised the need to develop an airfield, with hangars to store their aircraft. A 1937 meeting of the town committee, chaired by Bill Manson, proposed the present Luggate site as the only practicable option. World War II intervened, however, and nothing came of this proposal.

After the war an area of land on the Wanaka Station, at the southern foot of Mount Iron, was used as an airstrip. When its owner Mrs G.C. MacPherson, daughter of P.R. (later Sir Percy) Sargood, offered the strip for municipal use, the council declined. The airstrip was, however, used throughout the 1950s, 1960s and 1970s by a number of operators, including Jack Young, the Wanaka Aero Club and Luggate Game Packers. The first Wanaka-based commercial aircraft operation was set up in the early 1970s by the late Peter Plew and administered by the Southern Districts Aero Club, Gore. Peter later bought this company, giving it the name Aspiring Air Ltd, and operated a de Havilland Beaver, an aircraft ideally suited to the conditions. The airstrip was later sold to Sir Clifford Skeggs, who sealed it and operates it privately.

In 1965 the present Luggate location was bought as a farm by a Southland farmer and aviation enthusiast, John Allison. He farmed it for a short period and held the occasional aircraft fly-in. Norman Pittaway purchased the property on 14 May 1969 and has farmed it since.

Left An aerial shot taken on the afternoon of the 1992 airshow.

On 12 July 1971 Arthur Scaife and Jack Scurr of the Lake County Council met with representatives of Vincent County. The central point of discussion was that Wanaka needed an airport and could not rely on Alexandra or Queenstown, even if they were only one and a half hours down the road! Years of negotiation between the county representatives followed. A key figure throughout the 1970s was Lloyd Dunn, a pilot and later Chairman of the Wanaka Airport Authority. He kept the negotiations going and pressured the authorities. Other key figures in the discussions were Stan Kane (Chairman of Vincent County Council), Tommy Thomson (Chairman of the Lake County Council), Bill King, Maurice Duckmanton, Jim Kyles and Neil Studholme. In the late 1970s the Minister of Aviation in Wellington was visited by a deputation including Stan Kane and Tommy Thomson, and an application for a government subsidy was presented.

In 1978, with government support, 38 hectares were purchased from Norman Pittaway for $49,000, plus $8,000 for fencing. The government subsidy of $50,000, plus a contribution of $10,000 from the Vincent County and a loan of $60,000 were used for development of the land. Power, telephone and the sealing of a 400–metre strip were the priorities, at a cost of $120,000.

Right   Airshow day, 1988.

Three years later, in 1981, Alastair McMillan purchased Aspiring Air Ltd and in 1982 built a hangar on the Luggate site. A toilet block and water storage tank were also constructed. The 400–metre sealed runway was completed. That same year the Alpine Deer Group, under the management of Tim Wallis, set up its operations at the airport and constructed a hangar.

When the airport was officially opened by the Hon. Warren Cooper, Minister of Foreign Affairs, on 8 January 1983, an airshow was held to celebrate this major achievement. In 1985 office extensions were built by the Alpine Deer Group for the two full-time staff, Tim Wallis and his secretary Nola Sims.

By 1986, the Lake County Council realised that if it wanted to retain the airport, then redevelopment was urgently required. The runway was lengthened to 1,390 metres and sown in grass seed but this grew poorly because of the extreme weather conditions in the region. As a result only limited use was made of the airport. The council made the commitment to seal 1,200 metres of the runway.

Big money was needed and the committee that had worked so hard in the past took on the new challenge. Loans totalling $700,000 were raised — $250,000 from the Queenstown-Lakes District Council, $200,000 from the Wanaka business community, Air New Zealand $125,000 and the Ansett-Newmans Group

Left The same shot in 1990. The development of the airfield is evident when comparison is made with the 1992 shot on page 18.

$125,000. The government, however, showed a disappointing lack of interest. In 1987 the sealing was completed and Wanaka had an airport.

In 1988 the Alpine Deer Group held the first of their successful airshows, Warbirds on Parade; 14,000 attended. Over Easter Weekend of the following year, the RNZAF conducted 'Wise Owl 48' pilot and aircrew training exercise, and the Warbirds Association held a fly-in, watched by 3,000 people. A new privately owned hangar was constructed to house four aircraft.

In 1990 the Warbirds Over Wanaka show drew 30,000 spectators. That year ownership of the airport was transferred to the Queenstown Airport Corporation.

In 1992 a new hangar was constructed by the Alpine Deer Group for its new subsidiary company, the Alpine Fighter Collection. The final coat of paint was applied in the week before the airshow. In the same year permanent fencing was erected along the runway, the taxiways and the boundaries, and in March Carl Thompson erected a building and began operations as Wanaka Aviation, flying two Cessna aircraft on scenic flights to Milford, Mount Aspiring and Mount Cook. A hangar is to be constructed. On 18 and 19 April, Warbirds Over Wanaka attracted a crowd of 55,000. On 3 December Biplane Adventures Ltd was set up by Grant Bisset, offering aerobatic thrills and joy rides in a Pitts Special and a Tiger Moth. During the year an automatic meteorological station was erected and connected directly to Wellington. It is not used locally.

On 3–4 April 1993, the RNZAF held 'Wise Owl 60', a pilot and aircrew training exercise, and the following weekend, 10–11 April, the New Zealand Fighter Pilots Museum was opened. The Alpine Fighter Collection announced plans to build a third hangar. And Mike Kelly and Murray Patterson of Southair Aviation are to build a hangar to house their ex-Polish Air Force MiG 15.

Aspiring Air Ltd, currently part-owned by Alastair Mc-Millan and Barrie McHaffie and employing four full-time pilots, one office staff and several part-time pilots, operates two Britten Norman Islanders and four Cessnas. Scenic and charter flights as well as pilot training are offered. Graeme Knauf of Great Lakes Aviation Ltd, who has rented space in Aspiring Air's hangar for some years and operates as an aircraft engin-

Left A section of the crowd line in 1988. Fencing was constructed for the 1992 display.

eer with a full-time assistant, plans to construct a hangar. The Alpine Deer Group and its subsidiary the Alpine Fighter Collection currently employ five full-time and one part-time staff. At least two other privately owned hangars are in the planning stages.

Another exciting project involves the construction of a transport museum on neighbouring land. Gerald Rhodes, a long-time collector of historic vehicles and aircraft, has acquired about 6.5 hectares of land from Norman Pittaway. The collection, which includes over 60 vehicles, boasts an original World War I ambulance, a World War II Centurion tank, field guns, fire tenders, cars, and a large number of motorcycles. The growing aircraft collection, which currently stands at eight, includes an airworthy de Havilland Devon (ZK-RNG) and, on static display, a 1933 Flying Flea and an ex-RAAF Canberra bomber. A workshop will be set up to restore the aircraft and vehicles.

Warbirds On Parade 1988

The airshow was originally set down for 10 January, but was rescheduled to Saturday, 2 April. January is typically a busy time for tourism and the Wanaka business community felt that Easter, when fewer visitors were in the area, was a better date.

The support of the warbirds fraternity was revealed at its best for this airshow. When the Dragon Rapide aircraft ZK-AKU, originally scheduled to participate, was unable to make the journey, Tim Wallis rang Tom Williams, then owner of another Dragon Rapide, ZK-AKY, at 10 o'clock on the Friday night before the airshow. Williams phoned back at 11.30 pm confirming that he could make it. The Dragon Rapide arrived from Masterton at 12.30 pm the next day.

After the airshow the RNZAF Red Checkers display team was grounded as a fuel-saving measure, although this was, fortunately, only temporary.

The list of acknowledgements in the souvenir programme included an interesting paragraph: 'Should the combination of civil and warbird aircraft, vintage and veteran farm machinery and equipment be a success, it is hoped that it will become an annual or biennial event. Wanaka is a great venue for such an occasion.' Discussion with the Wanaka business community and the resulting vote confirmed the airshow as a biennial event.

For hospitality, ten out of ten to Prue and Tim Wallis. The weather was sunny but with strong westerly winds that raised the dust.

Left Cheryl 'Rusty' Butterworth atop the Tiger Moth, with Tony Renouf at the controls; 1988 airshow.

Above The sign on the hill says it all,
WARBIRDS '88 THANK TIM + PRUE.

Left John Denton, complete with 'Mae West', climbs down from ZK-VNM.

de Havilland Venom

The Venom was built under contract in Switzerland as a DH112 Mk FB.1 and served with the Swiss Air Force as an FB.50, J1634. It was acquired by Aces High Ltd, England, in February 1985, dismantled and shipped to New Zealand. Although packed in a specially prepared, humidity-controlled container, the aircraft suffered major impact damage during transit and almost two years of restoration work was required before it could fly in New Zealand. When acquired by Trevor Bland the Venom had accumulated only 47 hours since a total rebuild and was the first privately owned jet fighter in New Zealand.

ZK-VNM flew for the first time in New Zealand on 29 August 1987 in the colours of a Venom FB.1, WE434 'L' Lima allocated to Trevor Bland while serving with 14 Squadron out of Tengah, Singapore from 1955 to 1958. The RNZAF based in Singapore operated 16 Venoms under a lend-lease scheme with the RAF, until their return after three years of service in 1958. On 16 November 1991, with John Denton at the controls, the Venom suffered engine failure on take-off from Ardmore and crash-landed through a ploughed paddock and into a poplar

Above John Denton taxis the Venom out to the runway in preparation for his display.

Top right Under the afternoon sun Trevor 'TT' Bland and 'Mustang One-Five' lift the dust as they taxi close to the crowd line.

Bottom right Dakota ZK-DAK taxis in after its display.

Overleaf The Sea Fury's prop blades are frozen in time. The partially lowered flaps were the result of hydraulic problems.

plantation. The resulting damage was extensive, both to the aircraft and to John's pride. The aircraft was a write-off. John walked away from the crash and suffered the ignominy of a television interview describing the accident.

The Mk 1's registration was cancelled in November 1992 and a new Swiss-built Venom, a Mk 4 serialled J-1799 imported from the United States in mid-1992, took over the ZK-VNM registration the following month. The aircraft was test-flown by Grant Biel on 21 January 1993 off Ardmore and is expected to appear at Wanaka in 1994.

Fly-in 1989

It was Easter Weekend 1989 and the RNZAF had taken over the airport to conduct a seven-day training camp, 'Wise Owl 48', which involved trainee pilots and navigators flying eleven CT-4 Airtrainers, two Sioux helicopters and two F27 Friendships.

A Mustang, five Harvards, a Beaver and a Devon flew down from Ardmore for a holiday gathering of friends. Not realising this, 3,000 people turned up expecting to see an air display. This was a phenomenal number, considering there was very little advertising of the event. The crowd was not disappointed, however, as a flying display was quickly organised.

The Mustang did not confine its display to the runway area, but buzzed low over the crowd. An RNZAF Friendship displayed its air drop parachute technique by releasing a container onto its target from low level. The Roaring Forties Harvards, the Devon, the Beaver and the RNZAF's Airtrainers performed, joined by Alpine's Tiger Moth and Hughes 500. A memorable solo display by Flight Lieutenant Keith Adair in an RNZAF CT-4B excited the appreciative crowd.

Left The Devon NZ1808/ZK-KTT, with part-owner Barry Keay at the controls, lifts off to give its display.

Warbirds Over Wanaka 1990

In 1990, when Warbirds Over Wanaka was designated an official sesquicentennial event, the crowd was looking forward especially to seeing Stephen Grey flying a Spitfire. The star attraction, now resplendent in wartime colours, was making its first public appearance since an accident in January 1989.

On a sunny, warm day with a light wind but lots of dust, the 30,000 spectators enjoyed a very full programme of flying, which flowed smoothly and to schedule. This year air traffic control was used at the airfield for the first time; the Airways Corporation supplied a mobile control tower free of charge, with volunteer controllers on hand.

Memorable displays were given by Stephen Grey in the Spitfire, Trevor 'TT' Bland in the Mustang, John Denton in the Venom and Robbie Booth in the Sea Fury, now free from the hydraulic problems that had affected it in 1988.

Bryan Beck and Doug Maxwell gave two uniquely New Zealand displays of flying. Bryan, flying a Hughes 269C in an act entitled 'A Dog's Life', 'rounded up' a wool bale, with the help of the air stirred up by his rotor blades, and locked it in its pen. The second act saw Doug in a Robinson R22, with the help of shooter Mark Hutton, give a display of deer recovery on the hill behind the airfield.

The finale of this year's display was the 'battle' in which the airfield came under attack from a ground-based enemy invasion force. A Centurion tank was captured by the enemy and the defending air force was scrambled to chase them off. The enemy were beaten into submission by the continuous strafing and bomb runs of the aircraft.

The airshow was by now recognised internationally and among the crowd were overseas visitors, including 'Rusty' Leith from Australia, a former wartime pilot of Tim Wallis's Spitfire. A feature article appeared in the *Warbirds Worldwide* publication, large photos adorned the front pages of New Zea-

Previous page The sun rises on the Spitfire on airshow day. Note the disposable plastic cups stuffed up the exhaust pipes to prevent dust or debris being blown onto the engine valves.

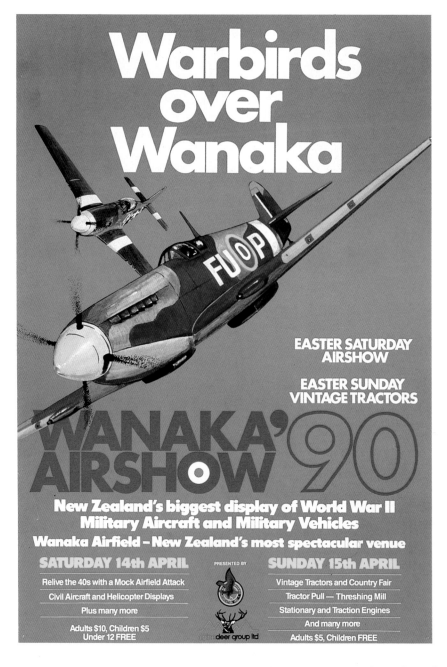

land newspapers and magazines, videos were produced and the flying sequences were shown on national television. Entrepreneurs discovered the show's marketability. The Wanaka community and surrounding districts benefitted to the tune of $1 million.

Contribution from Stephen Grey

Wanaka is this man's idea of paradise — especially when it fills with Warbirds and the Warbird fraternity.

The geography is unbeatable, the atmosphere electric and the hospitality outstanding.

I cannot speak too highly of the generosity of Tim Wallis, particularly in inviting me to fly his superb Spitfire at this unique airshow. What a privilege.

Split the jagged mountain horizon with that beautiful prop, gentle 'G' to the vertical, caress the ailerons and rotate the clouds. Sandwich still air with elliptical wings, singing Merlin, smiling faces. Magic — just magic.

Wanaka — keep flying.

Stephen Grey
The Fighter Collection

The Fighter Collection
This is the World War II derivative of the emblem used by the Escadrille Lafayette which flew in World War I. The original 1914–18 Lafayette emblem points the other way and has been modified over the years by its French Air Force successors. The Fighter Collection adopted the emblem as it signifies the first international co-operative fighter squadron with personnel and aircraft from various nations.
(Thanks to Stephen Grey)

de Havilland Devon/Dove

The de Havilland Devon/Dove was designed at the end of World War II as a replacement for the DH89. The prototype made its first flight on 25 September 1945, powered by two DH Gipsy Queen 70 engines.

From 1948 to 1981 the RNZAF operated 28 DH104 Devons and two DH104 Doves in the roles of communications, photographic, signal trainers and navigation trainers.

The aircraft were built by de Havilland to New Zealand specifications and differed from the RAF Devons. During construction there were no Air Ministry inspectors on the produc-

tion line and although arguably they could not be called Devons, New Zealand insisted. So 'Devon' they were called.

There are at least nineteen surviving ex-RNZAF Devon/Doves in New Zealand and overseas. Three are currently airworthy in New Zealand. ZK-KTT (NZ1808) and ZK-RNG (NZ1807) were operated as navigational trainers and ZK-UDO (NZ1821) flew in the RNZAF as a communications aircraft.

de Havilland Dragon Rapide/Dominie

The de Havilland 89B Mk.1 Dominie was built by de Havilland's sub-contractors Brush Coachworks Ltd, Loughborough, in Leicestershire, England in July 1943, c/n 6653.

The aircraft was handed over to No. 76 Maintenance Unit RAF as HG654 on 13 July 1943 and was shipped to New Zealand aboard the vessel *Port Chalmers* on 26 August 1943, arriving on 6 October 1943, and brought on charge at RNZAF Station Hobsonville, Auckland, as NZ525.

NZ525 served in communications and transport roles with the RNZAF until being demobbed and purchased by New Zealand National Airways Corporation (NZNAC) on 30 August 1946. Taking the civil identity Dragon Rapide and registered ZK-AKY, the aircraft (in keeping with NZNAC practice) was named after a native bird, 'Tui', and operated throughout New Zealand. The company later changed its name to National Airways Corporation (NAC).

Ritchie Air Services Ltd, operating from Lake Te Anau, acquired ZK-AKY on 19 May 1964 and flew the aircraft until 1969, when ownership changed to Tourist Air Travel. The aircraft changed hands again in 1972, this time flying tourists around Queenstown in the colours of Mount Cook Airlines. In 1978 ZK-AKY was retired and placed in storage at Queenstown.

Tom Williams of Te Parae, Masterton acquired the aircraft, along with a supply of spare parts, on 28 March 1978. A complete rebuild begun in 1983 concluded on 16 January 1986, when ZK-AKY re-emerged fully restored, wearing its old NAC colours, and made its first flight.

ZK-AKY was hired out to White Island Airways at Rotorua and flew tourist charter work during 1989–90. In 1991 the

Top left Dove NZ1821/ZK-UDO is fuelled from a former air force workhorse, a 1942 Thompson Aviation refueller which is part of the AFC Collection.
Bottom left The Warbirds Dakota receives some of the chief sponsor's product.
Overleaf A relaxed setting — Dragon Rapide, Harvard and DC-3.

Above Dragon Rapide ZK-AKY 'Tui' on arrival at Wanaka.

Croydon Aircraft Company at Mandeville (near Gore) leased the aircraft from Tom Williams for tourist charter work. In 1992 Gerald Grocott purchased ZK-AKY, which is now operated permanently by Croydon Air Services, Mandeville.

North American Mustang

ZK-TAF is a P-51D-30-NA model, c/n 44-74829, built in June 1945, and delivered to the USAAF the following month. In 1951 it was transferred to the RCAF as 9265 and sold in May 1959 as N8675E to Aero Enterprise of La Porte, Indiana. In the early 1960s it was owned by Dr Byram Burns of Marengo, Iowa, as N619M and N169MD 'Tangerine' and later as N769MD.

In 1968 the aircraft was damaged in a night landing incident and it was in this condition that Max Ramsay of Johnson, Kansas acquired N769MD in May 1976.

In the early 1980s the aircraft was owned by Fort Wayne Air Service, Fort Wayne, Indiana and Courtesy Aircraft, Rockford, Illinois, as N6175C. John Dilley of Muncie, Indiana purchased the still-damaged aircraft and commenced a rebuild using an

ex-Indonesian Air Force Cavalier airframe and modern avionics. It was during the rebuild, with the modern cockpit fittings already installed, that Ray Mulqueen acquired the aircraft for the Alpine Deer Group. Although the AFC philosophy is to have the aircraft rebuilt to as near the original specifications as today's regulations allow, so much work had been done on the Mustang that the decision was made to leave the modifications unchanged.

The aircraft was test-flown in November 1984 and imported by Tim Wallis the following month, securing the registration ZK-TAF (TAF for Territorial Air Force) in May 1986. On 24 January 1985 ZK-TAF was test-flown for the first time in New Zealand over Wigram by John Dilley.

The Mustang wears the colours of NZ2415, one of the more colourful Mustangs in service with RNZAF's TAF and flown on a regular basis by Squadron Leader Ray Archibald, the Commanding Officer of No. 3 Canterbury Squadron TAF. This connection accounts for the red and black colour scheme. The aircraft has a natural metal finish because there was little aerial opposition in the closing stages of the war, which meant that

Above left Arriving at Wanaka, Brian 'BJ' Rhodes, the ferry pilot, lifts himself out of the Mustang's cockpit. In the rear seat is Warbirds engineer Greg Ryan. The attention to historical detail is evident in the stencilling appearing on the aircraft, along with the No. 3 Canterbury Territorial Squadron crest:

NEW ZEALAND GOVT.
MODEL P-51D-30-NA
SERIAL NO. 44-74829

ROLLS-ROYCE V-1650-7
SEA LEVEL 1490 H.P.
SERVICE THIS AIRPLANE WITH
GRADE 100/130 FUEL

Above right Greg Ryan of Aero Technology Ltd, with rag in hand, soaks up engine oil from around the Merlin, notorious for leaking.

Above Stephen Grey taxis out to give the star display in the AFC's Spitfire, call sign 'Spitfire One-Six'.

Right On the Spitfire, the panel in the foreground is for access to the battery compartment and control cables. FU represents the squadron identity code of No. 453 (Australian) Squadron. P is the aircraft's individual identity letter. TB863 is the individual aircraft's RAF serial number. The upper surfaces of the fuselage and wings are dark green and ocean grey. The lower surfaces of fuselage and wings are medium sea grey.

any advantage Allied aircraft gained from using camouflage had been nullified.

The original NZ2415 was one of 167 P-51D-25-NT Mustangs ordered from the American government during 1945 in a lend-lease agreement. After Japan's unconditional surrender, however, orders for 137 were cancelled. But 30 Mustangs were in transit and could not be returned. NZ2415 was the fifteenth aircraft in the first batch of fifteen to arrive on 27 August 1945. The balance arrived in subsequent months and all were placed in storage at Hobsonville. NZ2415, along with the other 29, was eventually stored at Ardmore, still cocooned in its protective seal.

In April 1952, NZ2415 was assembled and allocated to No. 3 Canterbury Territorial Squadron, based at Wigram, as part of New Zealand's TAF. At some time during its service, the spinner was painted red and black, earning it the nickname 'Rudolph the Red-nosed Reindeer', but the authorities claimed that this colouring mesmerised ground crews (perhaps it was a case of individuality being frowned upon) and it reverted to silver.

On 23 January 1954, NZ2415 was involved in a landing accident at Wigram, caused by engine failure. There was damage to the undercarriage doors and flaps and the upper main planes suffered wrinkles. The aircraft was sent to Woodbourne in May 1954 for repair, but this did not eventuate. On 2 May 1958, NZ2415, still in its damaged state, minus engine and propeller and only 286 hours logged, was sold, along with another damaged Mustang, NZ2410, to ANSA Ltd of Upper Moutere, Nelson for £50 and scrapped.

In order to finance the purchase of the Spitfire in 1986, the Alpine Deer Group put ZK-TAF up for sale and for a while it seemed that the aircraft would leave New Zealand. Fundraising ventures succeeded in keeping ZK-TAF in the country but it was not until the New Zealand Historic Aircraft Trust acquired the aircraft that its future was assured. In 1992 ownership changed to John Sager, an Air New Zealand 747 captain, of Auckland, and Graham Bethell, a Hong Kong-based 747 captain for Cathay Pacific.

In early 1992, ZK-TAF took on an advertising role for a Singapore beer company and was painted in a bright yellow

scheme complete with shark's teeth and eyes on the engine cowling. It carried the fuselage letters TIG-8 and, in place of the customary roundels, the fuselage and wings sported a growling tiger's head. The flying for the advertisement, which depicted an air race, was filmed at Wigram and in the skies over Queenstown. When the advertisement was completed the temporary decoration was peeled off.

Contrary to popular rumour, the flaking yellow paint (said to have been ingested into the engine) was not the reason for the aircraft being grounded at Dunedin before the 1992 airshow. Internal corrosion of the seal for the cylinder head and block had resulted in coolant loss. This was discovered at Christchurch Airport on the flight south but the decision was made to continue the journey to Dunedin.

On shutdown at Dunedin the aircraft suffered a complete loss of coolant and there it remained for eight months until a replacement zero-houred engine arrived from the United States. The Merlin was installed, tested and then the Mustang made a quick flight to Auckland for the November 1992 Air Expo. Ironically, coolant system corrosion was one of the problems that eventually led to the TAF Mustangs being grounded and retired in May 1955.

North American Trojan

The T-28 Trojan was built in December 1956, as No. 226-140, Bu. No. 140563, and served with the United States Navy as an aircraft carrier training aircraft. It has a tail arrestor-hook, shortened propeller blades and wears the colours of VA 122 of the Pacific Training fleet based at Lemoore, California.

Retired in the mid-1970s, the aircraft was placed in outside storage at the Pima desert aircraft storage, Arizona. There it sat until being sold privately, reactivated, and made ready for resale. The aircraft was purchased by the late John Greenstreet in April 1989 and test-flown the following October from Ardmore. Registered as ZK-JGS, it is currently owned by the New Zealand Warbirds Syndicate based at Ardmore. An engine overhaul prevented the aircraft from attending the 1992 airshow.

This aircraft can outclimb the Mustang and Vampire to 10,000 feet at a rate of 4,200 fpm.

Above The mascot atop the instrument panel in the Trojan ZK-JGS belonged to the late John Greenstreet and sits there in remembrance of the warbird enthusiast. On the engine cowlings appears a yellow Trojan's helmet. The black striping down the sides of the fuselage follows the natural flow of the exhaust and hides the stains.

Top left Throttled back, canopy open, undercarriage down and locked, flaps and tail arrestor-hook extended, the only thing missing is an aircraft carrier. A display by ZK-JGS (JGS for John Greenstreet), flown by Keith Skilling.

Bottom left Trevor Bland, with a young passenger in the back seat, taxis the Mustang for a joy ride.

1990 NEW ZEALAND
TOURISM AWARD
COMMENDATION WINNER

CATEGORY: SPECIAL EVENTS
WARBIRDS OVER WANAKA

HON FRAN WILDE
MINISTER OF TOURISM

TOURISM
AWARD

Top The wings of the dreidecker ZK-FOK give a venetian blind effect.
Above The 1990 airshow won a New Zealand Tourism Award in the 'Special Events' category.
Right John Kelly in a Pitts Special taxis in on Good Friday, ahead of a constant stream of arrivals.

Redfern Fokker Triplane Replica

ZK-FOK is a full-size replica of the famous dreidecker (three decks). It was built by Stuart Tantrum working with John and Victoria Lanham at Levin, and involved 4,000 hours of hard work over a period of three years. The aircraft made its first flight on 23 February 1985.

Extensive research into a genuine colour scheme resulted in the choice of the war paint worn by Fok. Dr.1, 107/17. This aircraft was delivered in October 1917 from Schwerin and powered by a 110 hp Le Rhone engine operated with Baron Manfred von Richthofen's Jagdeschwader 1 from October 1917 until retirement in April 1918. Initially successful, the aircraft were popularised by the extrovert style of Richthofen and his 'flying circus'. But, as new technology increased combat altitudes, the pilots discovered that their aircraft were slow and underpowered, and that the drag from the triplane wing system deprived them of the competitive edge.

Rather than the familiar bright red of the 'Red Baron', the scheme chosen is a lozenge pattern worn by the World War I fighters of the German Army Air Service. Five colours are incorporated in the camouflage scheme: green, purple, ochre, prussian and blue.

The Fok. Dr.1, the first of its type to fly in New Zealand, is powered by a 220 hp Continental radial. It is currently operated by New Zealand Warbirds Syndicate out of Ardmore and has made the long trip south once to appear at the 1990 airshow, performing a dogfight sequence with Tom Grant's S.E.5a.

Warbirds Over Wanaka 1992

The Preparations

Early in the planning stages the organisers looked at how they could make this airshow better than the 1990 event. After discussions with Ray and Mark Hanna of the Old Flying Machine Company (OFMC) in England, sponsorship was pursued and the OFMC's Messerschmitt 109J was prepared for shipment to New Zealand to participate in Warbirds Over Wanaka 1992.

The costs of this undertaking were underwritten through offering 150 Messerschmitt 'Gold Pass' tickets to the public at $100 each. Although new to New Zealand, this is a common method of financing used at overseas airshows. A special enclosure and facilities are offered to these enthusiasts. The final crowd of 55,000 was testament to the high profile of the airshow.

The organising committee of Tim Wallis (chairman), his brother George Wallis (deputy chairman) and Gavin Johnston were the airshow co-ordinators for the three key areas of the weekend. Tim was responsible for the sponsorship and air displays, George for the vintage machinery display and Gavin for the trade stalls, car parking, and liaison with the Wanaka business community.

Tim was assisted with the civil air display co-ordination by Phil Murray and Simon Spencer-Bower. Phil is a pilot with over 5,000 hours of experience recorded in his log books. At the age of 25 he left behind four years of agricultural work, having amassed 2,700 hours, which included numerous airshow displays in his Midget Mustang, and signed up for a tour of duty with the RNZAF. He was taught to fly the military way. Currently a member of the Red Checkers, he is rated to fly the AFC's Zeke and Avenger. Simon Spencer-Bower is the owner/chief pilot of Canterbury Helicopters Ltd and has many years of flying behind him. He currently owns Tiger Moth ZK-BUO

Previous page Spitfire and
Messerschmitt in formation harmony.

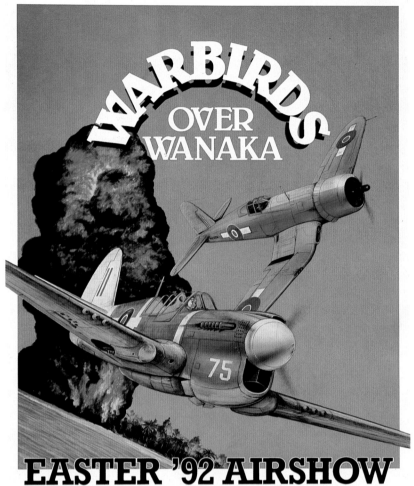

EASTER '92 AIRSHOW

New Zealand's biggest display of World War II
Military Aircraft and Military Vehicles

Wanaka Airfield — New Zealand's most spectacular venue

SATURDAY, 18TH APRIL — AIRSHOW —		SUNDAY, 19TH APRIL — VINTAGE TRACTORS —
Spectacular mock airfield **attack**		Vintage Tractors and Country Fair
Civil Aircraft and Helicopter Displays		Tractor Pull, Threshing Mill and **Joy Rides**
Full action packed day commencing at 10 a.m.	ALPINE FIGHTER MUSEUM	Stationary and Traction Engines
		A Family Day commencing at 10 a.m.
Adults $10, Children $5, Under 12 FREE		Adults $5, Children FREE

and Chipmunk ZK-DUC. Simon is rated to fly the AFC's Zeke
and will fly the Yak 3 when it arrives.

In late February 1992 Tim, Phil and Simon got together to
plan the civil display content. In deciding what civil aircraft
were to be invited to fly in the displays, a list of 22 possible acts

Left The Warbirds Over Wanaka organising committee: Gavin Johnston, Tim Wallis and George Wallis.

was drawn up. Each was given a ranking out of ten, with audience interest one of the key criteria. From this list thirteen aerial acts flew on show day.

The pilots were given approval by the Civil Aviation Authority to fly in the displays in their respective aircraft. The soloists were required to submit sketches of their acts. Formation positions were decided upon for the Cessnas, de Havillands and helicopters that participated in the opening fly-pasts. Alastair McMillan was flying controller for the display. The RNZAF's one-hour display at lunch time was directed by Squadron Leader Gordon C. Alexander of Support Group Headquarters, Wigram. The flying acts were the responsibility of each pilot.

The afternoon display of the warbirds was co-ordinated by John Lamont, the chief pilot for the AFC. John served with the RNZAF for twelve years, three as a radio tradesman and nine flying and instructing on Harvards and Devons. He also trained on Sioux and Iroquois and has 25 hours on Strikemasters. The winner of the coveted RNZAF Sword of Honour and the De Lange Trophy — it is very rare for one person to win both awards — John was invited in July 1993 to fly for the Old Flying Machine Company at the Duxford Airshow. He has almost 11,000 hours recorded in his log books.

The Merton Sword of Honour

The Merton Sword of Honour, presented by Air Vice-Marshal Merton, is awarded to the best pilot, navigator or air electronics officer of the *ab initio* course at Wigram.

The De Lange Trophy

The De Lange Trophy is awarded to the pilot who achieves the highest flying marks on a pilot's course.

53

The display was organised on the basis of aircraft and pilot availability. Solo acts were allocated six to eight minutes, the Roaring Forties ten minutes and the Spitfire/Messerschmitt dogfight as long as the pilots wanted. The three and a half hour programme saw the aircraft grouped by type, role and theatre of war in which they served.

The briefing at 9 am on airshow day was run by Air New Zealand pilot Chris Lee, director of flying for the weekend. He discussed the flying rules, which included no aerobatics below 500 feet and no fly-pasts lower than 100 feet. Ray and Mark Hanna of the Old Flying Machine Company had clearance to display their Messerschmitt down to 100 feet. If you did not attend the briefing, your act in the display was cancelled.

Temporary New Zealand private pilot licences were arranged for Ray and Mark Hanna. Sponsorship for the aircraft was arranged to allow the mass migration of warbirds to fly into Wanaka at minimal cost, especially the 30 or so aircraft that flew down from the North Island. It was to the credit of the organisers that the range of aircraft assembled allowed for an unbroken programme when invited aircraft failed to make the journey.

A small number of invited aircraft failed to arrive. These included the Mustang ZK-TAF (see p. 42) and Harvard ZK-MJN (NZ1052), which ground looped at Ardmore three weeks before the airshow. Damage was minimal and it was made airworthy within a short time.

The Trojan, ZK-JGS, was unavailable because of an engine rebuild. A Lockheed L-18 Lodestar, N56LH, from Australia suffered engine problems on its trans-Tasman journey and landed on one engine at Kaitaia. The passengers flew down to Wanaka in another aircraft. A Lockheed Neptune suffered a con-rod failure in Australia a week before the airshow, and on its flight south Tiger Moth ZK-CDU (wing-walker display) made a precautionary landing into soft sand on a beach near Kaikoura, smashing its propeller.

An intrepid group from Norway wanted to bring their twin-engined warbird to the show for the cost of the fuel. The Douglas Invader is based with the Scandinavian Historic Flight, Oslo, Norway. Sponsorship efforts were, however, concentrated on the Messerschmitt.

Above Glycol can be seen pooling between the wheels of the Mustang (background) at Momona Airport, Dunedin.
Left Dennis Egerton, ZK-HWH, and Tim Wallis, ZK-HOT, display the Hughes 500D.

The Messerschmitt Arrives!

A major drawcard for the 1992 Wanaka Airshow was the appearance of a Messerschmitt 109J. The aircraft, owned by Ray Hanna of the Old Flying Machine Company, is based at Duxford, the former World War II airfield in Cambridge, England. Ray, an expatriate New Zealander, learnt to fly in his native land in 1948 and then sailed to England to join the RAF. In 1966 his expertise at display flying led him to take over the highly prestigious position of leader for the RAF's Red Arrows display team. To date, his log books total over 17,000 hours flown.

The second member of the display team was Ray's son Mark, who was taught to fly by his father at the age of sixteen. Twelve years of service followed, flying fighters for the RAF. In the late 1980s he left the RAF and joined the OFMC as a director. More recently he has taken over the position of managing director. Mark has logged over 3,000 hours of flying time.

The Messerschmitt was shipped from England especially for

Above A smile from Mark Hanna. Note the impression the mask has left on his face.

Top left A rare view of the Messerschmitt as it waits in the Motor Holdings hangar to be reassembled. Note the spare wheel in the starboard wing undercarriage bay.

Bottom left A quick spot check by Roger Shepherd after a successful first flight.

The Old Flying Machine Company
The badge is that of the Flying Tigers, a squadron of American fighter pilots flying from bases in China against the Japanese during the late 1930s and early 1940s before the United States entered the war.

Contribution from Mark Hanna

The invitation to the Old Flying Machine Company to come to Wanaka at Easter 1992 was the beginning of an unforgettable New Zealand experience. It was with great anticipation that Roger, our Chief Engineer, and I arrived in Dunedin. Ray, my father, was to join us later on in Wanaka.

We remember well the outstanding hospitality offered by Prue and Tim Wallis. The support and generosity of other aircraft enthusiasts formed the basis of many new friendships.

The atmosphere generated by the '50,000+' people at the airshow reflects the high level of warbird interest in New Zealand. Most warbird fanatics will remember the dogfight between Ray in the Spitfire and me in the Messerschmitt. For me the highlight of the weekend was the opportunity to fly the newly rebuilt Kittyhawk. Ray Mulqueen, Tony Ayers and the team can be proud of their achievement — I had a very, very nice time flying the Kittyhawk.

Both my father and I would very much like to be involved more in the New Zealand Warbird scene. Warbirds Over Wanaka was a pleasure that we can only describe as pure fun. Thanks go to the Warbird Fraternity and to Gerard for this opportunity to be recorded in New Zealand's aviation history.

Mark Hanna

Mark Hanna
Managing Director
The Old Flying Machine Company

the airshow in a carefully prepared crate. It was the first time the 109 had been dismantled, as the aircraft flies to the airshows around Europe. In fact, two European airshow bookings were cancelled in order to make the New Zealand visit. British warbird lovers were initially concerned as they believed the AFC had purchased the aircraft.

Following its arrival at Port Chalmers the aircraft was trucked the remaining kilometres to Momona Airport, Dunedin. The OFMC's chief engineer, Roger Shepherd, reassembled the aircraft in Motor Holdings (Aviation) Ltd's hangar. During the first few days of work visitors were welcomed, but numbers increased to such an extent that those wishing to view the aircraft were required to make appointments.

For the small crowd gathered on the sunny Tuesday afternoon of 14 April 1992, the sight of the assembled German fighter parked alongside the Alpine Fighter Collection's Spitfire stirred many different emotions. To the returned servicemen these aircraft represented former battles and difficult times. But for younger enthusiasts these aircraft were knights of the air, examples of aerial technology that was at a peak during World War II.

With Mark Hanna at the controls, the first German aircraft wearing a swastika took to New Zealand skies for a short test-flight, concluding with a low, high-speed pass. On the aircraft's second flight Mark was joined by the AFC's Spitfire, piloted by Tim Wallis, and the Corsair, flown by Tom Middleton. The three aircraft gave a promotional fly-past over Dunedin and then flew on to Invercargill via Balclutha and Gore. The Spitfire and Corsair spent about twenty minutes over Invercargill while the Messerschmitt refuelled. The three fighters flew on to Queenstown before landing at Wanaka.

The Civil Aviation Display

A wide variety of civil aircraft performed in the morning.

Grumman Ag-Cat

The big yellow biplane took time off from its Queenstown Pionair Adventures to show its handling capabilities. Built in 1970, the Ag-Cat c/n 738 operated in the United States as N6598, a cotton sprayer, for about fifteen years and then found its way to Australia, where it was remanufactured and towed banners out of Brisbane as VH-JBB. The aircraft arrived in New Zealand in November 1991 and was converted to its present

role of carrying two passengers sitting side by side in the forward cockpit. What is offered is the open cockpit experience with leather jacket, flying helmet, goggles and obligatory scarf blowing in the wind.

Altavia Wilga

When imported by Dougal Dallison, the Polish-built Altavia PZL-104 Wilga (pronounced 'villga', meaning thrush) was the first of its type in New Zealand. Appropriately registered ZK-PZL, the aircraft made its first flight in New Zealand skies on 22 November 1991.

The aircraft's STOL (short take-off and landing) capabilities — 80 metres for take-off and on landing a ground roll of 60 metres — and flying speed of 37–100 knots allow it to include in its repertoire glider towing, topdressing, parachuting, surveillance/observation work, air ambulance and floatplane ops, to name just a few.

The RNZAF Display

The aircraft on display in 1992 included the CT-4B Airtrainers flown by the Red Checkers, an Iroquois, two Strikemasters, a Friendship, an Andover and a 'Macchi'. The Kiwi Blue parachute team also performed. (The appendices give a full list of personnel.)

The Red Checkers team can trace its display routines back to the late 1940s, when they were flown by instructors from the Central Flying School (CFS), at Base Wigram. Known simply as CFS Wigram, their Harvards performed a quartet display with wing tips tied together. Displays of three to five aircraft performed throughout New Zealand until 1967, when, at Base Wigram's fiftieth anniversary, the Harvards appeared for the first time with red and white checkered engine cowlings and the new name the Red Checkers.

The now familiar display of a formation of four aircraft and the solo was first adopted in 1967 and progressed yearly, with new manoeuvres being added. The display smoke-generating equipment was fitted to the aircraft in the early 1960s. In 1970

Top left Wilga ZK-PZL at rest.
Bottom left Phil Maguire shows his skills in the Ag-Cat.

Top Support crew refuelling Airtrainer NZ1938. Left to right, Sergeant Trevor Wood, Corporal Glenn Hughes, Flight Lieutenant Theresa Cunningham (Flight Commander PTS), and Leading Aircraftman Blair Hopkins.
Above Flight Lieutenant Phil Murray, who earlier in the day gave the 'crazy clown act'.

the colour scheme of mainplane grey for wings and fuselage and international orange for the whole of the rear fuselage, tailplanes, fin, rudder and wingtips, was adopted. The red and white checkers were enlarged and could now be seen from greater distances.

But the worldwide oil crisis in 1973 meant that the hangar doors remained closed and the Harvards never flew again in the distinctive colours. In 1977, with over 30 years of service above the clouds, the aircraft were retired. In 1976 a new breed of trainer had entered the hangar. The New Zealand built CT-4B Airtrainer, with its 210 hp engine, was considered under-powered in comparison with the Harvard's 550 hp, but its flying strengths were identified and the Red Checkers' routines were revived for a new generation. Gone were the tied wing-tips, to be replaced by the new mirror formation, where one air-craft flies inverted above a second, mirroring its flight.

In February 1993, the closure of Base Wigram was announced and the Central Flying School moved to Base Ohakea in July 1993.

The 1992 team was led by Squadron Leader Ian McClelland, Officer Commanding CFS, Red 1. 'Mac' joined the RNZAF in 1976, completing the last RNZAF 'wings' course to fly Harvards. He has gone on to complete over 5,400 military flying hours in a variety of aircraft in extreme conditions, from the cold of Antarctica to the heat of the Middle East.

Red 2 was Squadron Leader Hoani Hipango. 'Hippo' has accumulated over 2,700 flying hours, which include nine years' service with the Royal New Zealand Navy and a tour with the Fleet Air Arm of the Royal Navy. In 1989 he completed an instructor's course in the RNZAF and has served at PTS (Pilot Training Squadron, instructing student pilots at the initial stage of their wings course) and CFS. He has had two seasons with the Red Checkers.

Red 3 was Flight Lieutenant Nick Osborne. 'Oz' has logged over 3,200 hours since joining the RNZAF in 1981 and has flown Strikemasters and Skyhawks. In 1986 he was a member of the Kiwi Red Aerobatic team flying Skyhawks. He has had a tour of duty with the RAAF and 1992 was his second year with the Red Checkers.

Red 4 was airshow organising committee member Flight

Lieutenant Philip Murray. Phil joined the RNZAF in 1984 and had four years on Iroquois before turning his talents toward instructing in 1990.

Red 5 (solo aerobatics) was Flight Lieutenant Mike Tomoana. 'Tux' joined the RNZAF in 1984 and has over 2,700 hours logged on the Cessna Golden Eagle VIP transport aircraft and the P-3K Orion. In 1992 he took a position as instructor at PTS Wigram and performed as the soloist in his first year with the Red Checkers.

Red 6 (reserve) was Flight Lieutenant Peter Walters. 'Pee Dub', who joined the RNZAF in 1982, has over 3,100 hours logged, predominantly on helicopters. After pilot training he converted to the Iroquois flying operations in Singapore. In 1990 he completed the instructor's course and is currently an instructor on helicopters. He gave the commentary at the 1992 airshow.

Above The Red Checkers performing their Vic loop.

Above Strikemaster NZ6363 flown by Flight Lieutenant Mike Longstaff, an RAF exchange pilot, lifts off in preparation for its display. The Strikemaster made its last appearance at the airshow as an operational RNZAF jet pilot trainer. Retirement in 1992 ended twenty years of service of the type with No. 14 Squadron at RNZAF Base Ohakea. Sixteen were operated in the role of advanced jet training but the discovery of fatigue cracks throughout the Strikemaster fleet led to their retirement. Two were lost in crashes, seven have been put up for sale (including NZ6363), six have become RNZAF instructional airframes and the last has been allocated to the RNZAF Museum.

Top left Andover NZ7620, flown by Flight Lieutenant Neil Kenny of No. 42 Squadron, RNZAF Base Auckland, makes a slow pass displaying its parachute drogue drop technique.
Bottom left Iroquois NZ3805 of No. 3 Squadron Detachment RNZAF Base Wigram. In the cockpit are Flight Lieutenant Mark Cook and Flying Officer Anthony Fesche, while Flight Sergeant HCM (helicopter crewman) Terry Houghton sits by the open door.

Pre-World War II Aircraft

de Havilland Moth

The DH60M Moth was introduced in 1928 in answer to the need for a sporting light aircraft. Its 100 hp Gipsy I engine led to the type becoming known informally as the Gipsy Moth. Constructed with a steel-tube fuselage (hence the M for metal), it was capable of a maximum speed of 88 knots.

New Zealand's sole airworthy example (and incidentally its oldest airworthy aircraft) was specially modified on the production line as a racing machine for de Havilland's chairman, Alan Samuel Butler, on 26 June 1930. Registered G-AAXG, and powered by a 120 hp Gipsy II engine, the aircraft has an exciting racing history which began just nine days later, on 5 July 1930, when it entered the prestigious King's Cup. It was flown into second place by Alan Butler, at an average speed of 130 mph. The Moth crossed the channel and, flown once again by Alan Butler, won the Round Europe Challenge de Tourisme Internationale. In August 1930 the aircraft was sold to a prominent Frenchman Edouard A. Bret as F-AJZB and the following month won the 1930 Coupe Zenith Internationale, averaging 112 mph for the 1,036 miles around France. Pilot and aircraft were successful the following year in the same race.

In February 1933 the Moth returned to England and into the ownership of Brian Lewis & Co. Ltd. It was re-registered G-AAXG. The aircraft was sold in June 1933 to Sub-Lieutenant H.R.A. Kidston, Royal Navy, who brought the aircraft, along with his Bentley, to New Zealand on HMS *Diomede* as part of his personal luggage. In August 1933 the aircraft was unpacked at Hobsonville, New Zealand.

The Moth was purchased by R.J. Tappenden on 13 October 1935 and registered ZK-AEJ, receiving its New Zealand certificate of airworthiness on 3 March 1936. Before this the aircraft had an Air Ministry certificate. Jack Allen of Bunnythorpe, Hawke's Bay, purchased AEJ in December 1935. On 9 September 1939 it was impressed into RNZAF service but was stored and not given a serial number. In June 1940 the aircraft was returned to its owner but did not fly again until 20 December

1945. In 1947 R.N. Brown of Hamilton bought the Moth and flew her until 1 August 1950 when Jack Allen repurchased her. The aircraft was sold to Gordon Reader on 12 September 1966 and, after fifteen years of storage, ZK-AEJ was bought by Gerald Grocott, a Kiwi flying for Swissair. The aircraft was rebuilt in 1987 by New Zealand's 'Mr Moth', the late Temple Martin, and his son Gary, at Hastings' Bridge Pa Airfield, with Bob McGarry of Christchurch overhauling the Gipsy II engine. ZK-AEJ moved south to the Croydon Aircraft Company, Mandeville (near Gore) in 1990 after the North Cape to Bluff Moth Rally.

A feature of this aircraft is its folding wings, allowing for easy hangar storage. It wears an all silver scheme with the logo of the 1930 Touring Competition — Challenge International K5 1930 — appearing on both sides of the forward fuselage.

Isaacs Fury II

Based on the 1930 design, this seven-tenths replica of an Isaacs Fury II (illustrated on page 2), built by John H. Ross of Oamaru, first took to the skies in 1982, powered by a Rolls-Royce engine producing 100 hp.

Rex Carswell, making his third appearance, exemplified the extra commitment required from an owner of this type of aircraft in order to attend the airshow. Not for him a flight of a few hours, but a journey of two days, cruising at 70 knots. On the first day he flew Hobsonville – Raglan – Hawera – Foxton – Blenheim. On day two the Fury travelled Blenheim – Kaikoura – Rangiora – Timaru – Omarama – Wanaka. Then, of course, there is the return flight!

Throughout the journey, Rex was accompanied by a support aircraft with reserve fuel as some of the airfields lacked refuelling facilities. Regulations state that barrels of fuel cannot be carried within the aircraft, so the Fury was refuelled by pumping fuel from the support aircraft's own tanks.

The colour scheme worn by the replica is that of a Hawker Fury I flown by the RAF during the 1930s. Considered by many to be the most elegant of the RAF's biplane fighters, the Fury is recognised as the first RAF fighter in squadron service to exceed 200 mph.

Replica Plans S.E.5a

Built from Canadian plans over a two and a half year period, Tom Grant's S.E.5a (Scout Experimental No. 5a) took to the skies in July 1977. The aircraft was the first of its type to fly in New Zealand. The seven-tenths scale replica, powered by a 100 hp Continental engine, looks convincing with its replica Lewis machine gun mounted on the upper wing.

The original fighter aircraft on which the replica is based was designed by H.P. Folland and built by the Royal Aircraft Factory, which later changed its name to the Royal Aircraft Establishment (RAE) to prevent confusion with the initials of the Royal Air Force. The S.E.5a is considered one of the outstanding single-seat fighters of World War I.

The Pilot Trainers

de Havilland Tiger Moth

The Tiger Moth was designed in 1931 by the de Havilland team, led by technical director Captain Geoffrey de Havilland. The type owes its name to this man, known simply as DH, a fervent lepidopterist — a collector of moths.

For over 60 years New Zealand aviation enthusiasts have had a love affair with the Tiger Moth. The aircraft has a unique place in New Zealand aviation history as the most numerous type to be operated by both civil operators and the RNZAF. Some 477 examples appeared on the civil register between December 1937 and November 1956.

The RNZAF operated 335 of the type as 'ab initio' trainers from 1939 to 1956. The survivors were retired and now make up the majority of the civil listings. Surplus RAF and RAAF Tigers have also appeared on the New Zealand civil register.

At the outbreak of World War II Tiger Moths operating with aero clubs were impressed into service and construction began in earnest in 1939 at the de Havilland Aircraft Company's Rongotai factory (today's Air New Zealand Wellington terminal). The first aircraft were turned out before the factory had been completed. Some 181 Tigers were constructed on this site.

Top left Murray 'Mo' Gardner runs up the Gipsy II engine before giving his display.
Bottom left Tom Grant in his seven-tenths scale replica S.E.5a lifts off to give his display. This was the aircraft's third appearance at the airshow.

New Zealand's post-war aviation history records the Tiger Moth in the new role of topdresser. Makeshift hoppers were fitted to the front cockpits and the Tiger Moth pilots proved that the aviation industry had a role to play in New Zealand's farming future. Aero clubs also re-equipped themselves with the readily available and inexpensive peacetime pilot trainer.

Seven Tiger Moths were in attendance at the 1992 airshow. On 31 December that year New Zealand could boast 46 Tiger Moths on the civil register, 31 of which were airworthy.

North American Texan/Harvard

Retired in June 1977 after 36 years' service with the RNZAF, the AT.6 Harvard, with its rugged good looks and never-to-be forgotten rasping note from the propeller tips as they go supersonic, is seen by many as the epitome of the warbird classics. For the pilot the cabin noise is around 92 decibels so flying helmet and mask mike are essential, as is a flying suit because the Harvard is such an oily old warbird.

In March 1941, after lengthy delays, the first of the 105 Harvards originally allocated arrived and were used by the Empire Air Training Scheme for advanced pilot training. During their long RNZAF service the aircraft were employed by a number of squadrons and units in a multitude of roles.

Of the 202 Harvards that were eventually operated by the RNZAF, 53 are still extant in New Zealand and overseas. Thirteen were flying in New Zealand skies on 31 December 1992, and an ever growing number are being rebuilt to fly in the not too distant future.

One aim of the 1992 airshow was to bring together ten of 'the old birds'. On the day nine appeared and flew a unique diamond nine formation. To have so many privately owned Harvards flying together was a Southern Hemisphere first, a sight not seen in New Zealand since the type was put up for tender in 1978. (The appendices list the aircraft and owners.)

de Havilland Canada Chipmunk

The 'Chippies', as they are affectionately known, were designed by the Canadian branch of de Havilland at Toronto. The

Left Simon Spencer-Bower displays his Tiger Moth ZK-BUO/NZ795 '46'.
Below Ken Walker lifts ZK-ENC/ NZ1091 into the air. The Americans referred to the type as the Texan.

Right The Harvards flying a diamond nine formation. The aircraft are, top to bottom, left to right, NZ1092, NZ1037, NZ1078, NZ1033, NZ1098, NZ1065, NZ1091, NZ1066, and NZ1099.

prototype's first flight took place on 22 May 1946. Two were displayed at the 1992 airshow, ZK-MUH and ZK-TNR.

ZK-MUH was built in England in September 1952 and until 1968, registered as G-AMUH, served as a trainer for BOAC pilots at Hamble, Southampton. In 1968 it was sold to a private syndicate before being acquired by Dougal Dallison in 1985 and based at Cranwell. The aircraft arrived in New Zealand three years later and was re-registered ZK-MUH. It was later passed on to John Wall, a retired Air New Zealand captain who flew it for a short period. In 1991 it moved once again, this time into the hangar of the Alpine Fighter Collection, and was repainted in the authentic scheme of WB568, RUC-G, of the Cambridge University Air Squadron, RAF, circa 1976, in order to match the colour scheme of its sister ship ZK-TNR.

ZK-TNR was acquired by its current owners, the Auckland-based DH Chipmunk Syndicate, in 1988 and repainted in its 1952 RAF colour scheme of WB566 '59' when it was based at Sywell, Northamptonshire.

Above ZK-MUH, alias WB568, displays its attractive new colour scheme in the hands of Simon Spencer-Bower.

The Roaring Forties

Above Low orbit in box formation.

This section is dedicated to the memory of John Greenstreet, who was tragically killed on 25 February 1990 while flying a Harvard (NZ1025) at Ardmore, during a Roaring Forties practice session.

The aerobatic display team that evolved from a military background came together originally for some Sunday afternoon entertainment and quickly discovered a public keen to see aerobatics.

The original members of the Warbirds Display Team — Trevor Bland, Ernie Thompson and John Denton — were joined by Keith Skilling, John Lamont, John Greenstreet, Steve Taylor, Robbie Booth and John Peterson as newly restored Harvards came on-line. The name for the team, the Roaring Forties, was coined by its members to reflect New Zealand's latitude, as well as the age of its pilots and aircraft. The aerobatic flying team of five is highly sought after at New Zealand and overseas airshows.

In 1992 the five pilots making up the team were led by John Lamont. RNZAF trained and currently a captain for Air New Zealand on Boeing 767s, John is chief pilot for the AFC and currently holds a rating for the AFC's Kittyhawk and Corsair.

No. 2 was Steve Taylor, a company director in the motor vehicle industry who comes from a general aviation background. He has displayed New Zealand Warbird aircraft for six years, including the Sea Fury and Pitts S-2B.

No. 3 was Robbie Booth. Robbie, too, is a company director with a general aviation background. He is an original syndicate member of the Sea Fury, in which he puts on precision displays, and was recently rated to fly the Venom.

No. 4, John Peterson, RNZAF trained, is currently Boeing 737 fleet manager for Air New Zealand and is a captain on Boeing 737s. He also holds an A category RNZAF instructor's rating.

No. 5 (soloist) was Keith Skilling. RNZAF trained, Keith currently flies as a flight instructor for Air New Zealand on Boeing 737s. He is the chief flying instructor in the New Zealand

Left John Greenstreet.
Below Steve Taylor (NZ1078) is the picture of concentration during the team take-off. His attention is wholly on his leader off the port wing.

Right The soloist (NZ1099) coming out of the vertical 360° and stall turn.

Warbirds Association and has been a member of the Roaring Forties for six years. He is rated to fly the AFC's Corsair and Kittyhawk.

To bring these pilots together, especially those flying for Air New Zealand, required many months of advance planning to arrange their annual leave for the Easter break.

The group is very dedicated, their commitment reflected in the detailed briefings conducted both before and after displays. They performed in October 1991 before a crowd of 180,000 at the RAAF's 70th Anniversary Airshow, which was held at Richmond, Sydney. The team flew Harvards lent to them by Australian Warbird members. In November 1992, they performed before a crowd of 230,000 at Auckland's Air Expo '92.

Sport and Utility Aircraft

de Havilland Canada Beaver

From gun running to airshow performer, the de Havilland Canada Beaver appears to have done everything. On the South African register as ZS-DCG, the Beaver is reputed to have been used for gun running in northern Africa before coming to New Zealand, and a stint of topdressing at Masterton. The aircraft, which has been a member of the Warbird fleet since 1984, wears the bright red colours of an RNZAF Beaver that was used in the Trans Antarctic Expedition.

The original Beaver carried the serial NZ6001, which was subsequently changed to NZ6010 to prevent confusion with RNZAF's Gloster Meteor III, the original NZ6001.

The Beaver is powered by a 450 hp Pratt & Whitney R-985 supercharged engine.

Douglas C-47B/DC-3 Dakota

The C-47B was built in 1943 by Douglas Corporation at Tulsa, Oklahoma for the USAAF as a C-47B-10-DK under contract, AC-40-652, being allocated the USAAF serial number 43-49219. The name Dakota was given to the aircraft by the

Above Peter Rhodes lifts off to display ZK-CKH, 'City of Auckland.'

RAF. The constructor's number 15035 was later changed to 26480. The number currently worn on its fuselage is 26840 — a faux-pas?

In the 1950s the aircraft was operated by Philippine Airlines as PI-C486 before being sold to Papuan Air Transport, Port Moresby, in 1960. Between 1970 and 1984 the aircraft, now registered VH-PNM, was operated by Ansett Airlines, Bush Pilots Airways of Cairns and Air Queensland before doing museum duty at McKay Air Museum as VH-SBT and then coming to New Zealand in 1987.

The aircraft wears the olive drab and neutral gray (American spelling) with the black and white D Day invasion stripes of F-I2, a C-47 which was operated by 48 Squadron RAF and flown by New Zealander Squadron Leader Rex Daniels DFC during the D Day invasion in June 1944.

The WB appearing on the fuselage is a USAAF touch, standing for warbird. ZK-DAK is a popular warbird, offering enthusiasts the chance to climb aboard and experience a piece of aviation history. The serious aircraft spotter will recognise that the

ZK-DAK registration was once carried by a Cessna 177-200, which crashed in 1978 near Nelson and has since been reduced to spares. This is one of the few occasions on which a registration for fixed-wing aircraft has been reissued; it is not a standard practice, as 'old aircraft never die, they just get stored away for another day'.

Above Time to count the rivets. The Warbirds Dakota climbs into the air above Barry Harcourt's camera.

Cessna Bird Dog

The name Cessna has been synonymous with American aviation since 1911. In 1927 Clyde Vernon Cessna founded the

Above John Denton shows off the excellent handling characteristics of the Vietnam War veteran. The name Bird Dog aptly fits the L-19 as its mission was to seek out and mark enemy targets — similar to the way a bird dog searches out and marks a bird's position.

Cessna Aircraft Company that has, over the years, manufactured tens of thousands of aircraft. The Bird Dog was designed for the US Army Field Forces for use as a light reconnaissance, observation and liaison aircraft. The prototype first flew in January 1950.

The Cessna L-19A/0-1G Model 305 Bird Dog on display at the 1992 airshow was the first of its type to fly in New Zealand. It was built in 1953, c/n 22737, and served with the US Army, serialled 51-16903, before being passed on to the South Vietnamese Air Force. As the North Vietnamese forces advanced, its engine crank-case was smashed, making it unserviceable, and the aircraft was left at Bien Hoa airport.

The Bird Dog was imported from Vietnam by Paul Dodd for Raven Air Ltd in 1990 and assembled by Aero Technology at Ardmore under the supervision of Greg Ryan. ZK-FYA joined the New Zealand Civil Aircraft Register in August 1991, having made its first flight a month earlier at Ardmore. The current owners are Bird Dog Aviation, Ardmore.

The Dogfight:
Spitfire vs Messerschmitt

Vickers Supermarine Spitfire

When Reginald Mitchell, the Spitfire's designer, was told that Vickers had suggested to the Air Ministry that the aircraft be named Spitfire, he said it was 'just the sort of bloody silly name they would choose'. But if he had lived to see his creation fly into combat, with a name that was radical for its day, and seen the respect it received from the enemy and the hope it gave to a beleaguered nation, he may have reconsidered his statement.

The AFC's Spitfire, a type 380, c/n CBAF.10895, was manufactured by Vickers-Armstrong at their Castle Bromwich 'shadow factory', near Birmingham in late 1944. It was primarily intended for the low to medium altitude role and carried the

Above Ray and Mark Hanna, father and son. A last-minute discussion before they 'do battle'.

Mark designation LF XVIe — a low (altitude) fighter (LF), sixteen 'e' (XVIe). The sixteen indicates that the aircraft is the sixteenth mark of Spitfire to have been developed. The 'e' describes the type of wing fitted to the aircraft. It had a clipped wing configuration that enhanced manoeuvrability at low altitudes and an armament of only two 20-mm Hispano cannon, one in each wing. At this stage of the war the Allies controlled the air over western Europe and dogfights with the Luftwaffe were few. There was also provision for two .5-inch and four .303 Browning machine guns. These, however, were not installed. A 500-pound bomb was slung under the fuselage and two 250-pound bombs were carried under the wings.

The introduction of the first American, licence-built Rolls-Royce Packard Merlin M266 engines created a new mark of Spitfire, the Mk XVI. This differed from its English cousin the Merlin 66, installed in the type 361, Mk IX, which employed electro-pneumatically operated supercharger gear rather than the M266's electro-hydraulic system.

Wearing the standard colour scheme of the day and carrying the codes FU-P, TB863 was prepared for its first operational mission. This was on 24 March 1945 with No. 453 (Australian) Squadron at RAF Matlask, in Norfolk. The squadron was tasked with armed reconnaissance and rail interdiction missions against targets in Holland. Also high on the list of targets were the heavily defended V2 sites that were striking southern England. In the six weeks of operations before the war ended, TB863 had logged 23 hours and 55 minutes, flying twelve missions. Its last operational sortie was 12 May 1945.

On 3 May 1945, TB863 was one of twelve Spitfires that escorted a C-47 Dakota returning Queen Wilhelmina of Holland from her wartime exile in England.

In June 1945, TB863 moved briefly to No. 183 (Gold Coast) Squadron and then to No. 567 Squadron in July in the role of Anti-Aircraft Co-operation Unit (AACU), becoming a target for gunnery practice. In June 1946, TB863 was issued to No. 691 Squadron and continued to fly in the AACU role. In February 1950, 691 Squadron was renumbered No. 17 Squadron (RAF); the original 17 Squadron was disbanded the same month at Miho in Japan and reformed in Britain simply by renumbering it No. 691.

Left Ray Hanna speaks to air traffic control as he waits for clearance to taxi.
Below Ray Hanna makes a low pass after take-off as TB863's undercarriage retracts. There is a tendency for the aircraft to swing to the left on take-off, owing to the anti-clockwise direction of the propeller blades. The use of rudder to the right counters this.

In July 1950, TB863 made its acting debut. Painted in Luftwaffe colours, but retaining its 17 Squadron codes UT-D, it re-enacted for the Farnborough crowd the Luftwaffe's defence against the precision bombing attack by No. 487 (New Zealand) Squadron Mosquitos. The bombing raid on Amiens prison on 18 February 1944 succeeded in freeing French resistance fighters.

In February 1951, TB863 came under the control of No. 3 Civilian Anti-Aircraft Co-operation Unit at Exeter. In July 1951 the aircraft suffered a take-off mishap and, although classified as repairable, was struck off charge.

Metro-Goldwyn Mayer (MGM) purchased the crash-damaged aircraft for their 1955 film *Reach for the Sky*, which told the story of the famous World War II fighter pilot, Sir Douglas Bader. TB863 is believed to have been used in cockpit shots. At the end of filming MGM placed the aircraft in storage. In 1967 the covers were removed and an incomplete TB863 lost yet more of its parts to the Spitfires that flew in the epic film *Battle of Britain*.

With filming complete, the aircraft's ownership was passed into the hands of A.W. Francis, who moved TB863 to his home at Southend-on-Sea in Essex, in December 1968. Initial restoration work began and in 1972 the aircraft was placed on display at the Historic Aircraft Museum at Southend Airport. The intention to undertake a serious restoration programme saw the aircraft move to the Imperial War Museum's facilities at Duxford Airfield in July 1974. The planned work did not eventuate and in 1977 the aircraft returned to its owner, who was now living at Southam in Warwickshire.

In October 1982, Personal Plane Services at Booker were contracted to start the restoration and began work on the wings. The civil registration G-CDAN was allotted to the aircraft in November 1982, in the joint ownership of J. Parks and A. Francis. Stephen Grey became interested in the project and acquired the warbird in late 1985, moving it to Duxford, the base for The Fighter Collection. David Lees, with participation in five Spitfire restorations already to his name, now added number six to the list.

A year later the New Zealand connection began when Tim Wallis, looking for a Spitfire with a wartime record to add to his collection, approached Stephen Grey. Grey wanted to add a Bell P-63 Kingcobra to The Fighter Collection so a deal was done. Two New Zealand engineers joined the restoration team to gain relevant experience: Marnix Pyle arrived in August 1987, and was joined later by Don McInnes.

During 1987 it was announced that Tim had purchased TB863 from The Fighter Collection in England and that the aircraft would be flown out from England in the cargo hold of a Boeing 747 in time for the 1988 airshow. This, unfortunately, did not eventuate as the restoration was not completed.

The aircraft was test-flown by Stephen Grey on 14 September 1988 with a second flight on the fifteenth, the day used to commemorate the Battle of Britain. With ten hours of test flying successfully completed, the aircraft was then dismantled and carefully packed into a crate for its sea voyage to Port Chalmers, New Zealand. It arrived in late December 1988 and was trucked the remaining kilometres to its new home, Wanaka. A new era in New Zealand aviation began when the crate was opened in the first days of the new year.

When the Spitfire was meticulously reassembled, the task went to Ray Mulqueen, Alpine's chief engineer, to fire up the Merlin for the first time in New Zealand. With the engine runs completed, Stephen Grey stepped back into the picture, test-flying TB863 on Wednesday, 25 January 1989.

Stephen congratulated the Wanaka team on their excellent work and the following day flew the aircraft north to RNZAF Base Wigram in Christchurch, where Tom Middleton, an experienced corporate pilot, became the first present-day New Zealander to be converted onto the type. Stephen made a goodwill visit to RNZAF Base Ohakea near Palmerston North and then it was on to RNZAF Base Whenuapai in Auckland, where Tim undertook his conversion and entered a new type in his log book.

On Saturday, 28 January 1989, TB863 was displayed by Stephen Grey at the Warbirds' Ardmore airshow in Auckland. After the display, Tim flew a second conversion flight in preparation for the ferry flight to Wigram, where the aircraft was to be painted in its camouflage scheme. Tim, inexperienced in the finer points of fuel transfer between wing and fuselage tanks, encountered problems during the flight south later in the after-

noon. The Spitfire, which had been insured only days before, force-landed at Waipukurau in central Hawke's Bay. Tim walked away unharmed, placing all blame on himself.

The aircraft was substantially damaged, but it was repairable. The world was scoured for parts and the engine sent to the United States for examination. Fourteen months later, following a careful rebuild by the Blenheim company Safe Air Ltd, the aircraft was test-flown once again by Stephen Grey.

To the trained eye there was a different look to the aircraft. The replacement undercarriage had been designed for the later marks of Merlin Spitfire. The undercarriage axles were straight, as compared with the original splayed (turned outward) set for grass runway use. Take-offs and landings on the hard asphalt runway surfaces are possible without fear of excessive wear to the tyres.

Over Battle of Britain weekend 1990, to commemorate the fiftieth anniversary of the air battle between the Royal Air Force and Luftwaffe, Tim Wallis undertook a fourteen-hour marathon flight around New Zealand in TB863. The flight was spread over three days, beginning on Friday 14 September, and co-ordinated with RSA commemorative services held throughout the country. Apart from a gasket replacement at Nelson, it went off without a hitch.

On 18 November 1992 Tim was flying to Auckland for Air Expo '92 when the aircraft suffered its second mishap. He was landing at RNZAF Base Woodbourne, Blenheim for a scheduled fuel stop when, approximately two-thirds of the way down the grass airstrip, the aircraft was caught by a strong cross-wind gust. The undercarriage clipped a tarseal taxiway directly ahead, causing the port undercarriage leg to collapse in under the fuselage, and the Spitfire started to ground-loop, coming to rest on its port wing.

The damage, described as 'substantial', was repaired once again by Safe Air Ltd of Blenheim. The Merlin was pronounced okay following inspection, and new propeller blades were obtained from Germany and an undercarriage leg from England. Further problems were encountered, however, during engine tests in September 1993, and the engine was sent to the United States.

Left One of the most evocative engineering designs in human history. Today the curves are still inspirational.

Above At rest. Ironically, the Spitfire has German-made Hoffmann propeller blades whereas the Messerschmitt's blades are made in England. Also of interest is the Spitfire's gun camera port in the right wing root.

Messerschmitt 109J

The Messerschmitt 109J, originally a Bf 109G-2, was manufactured in Germany in 1943 and exported, unassembled, to Spain as a 109J, without engine, tail assembly or armament. Hispano Aviacion was given the manufacturing licence and initially fitted an Hispano Suiza V12 engine but later changed this for a Rolls-Royce Merlin, the aircraft becoming an HA-1112-M1L, c/n 170. It was dubbed Buchon (a high-breasted pigeon common to Seville), because of the deeper, more rotund nose resulting from the installation of the Merlin.

The 109 is believed to have flown with 7° Grupo de Caza-bombardeo of the Spanish Air Force, possibly in the Sahara. It was serialled C4K-107 and served from approximately 1954 to 1966 flying ground attack against guerillas. Its armament consisted of two 20-mm Hispano wing-mounted cannon with provision for eight 80-mm Oerlikon rockets.

In July 1966, C4K-107 was one of twenty-eight Buchons pur-

chased by Hamish Mahaddie for use in the film *Battle of Britain*, performing a taxiing role. It was subsequently sold to the late Earl Reinert and shipped to the Victory Air Museum, Illinois, by 1976. In 1986 it was registered to Gordon Plaskett as N170BG and restoration began.

The 109 returned to England and on 6 May 1988, following an extensive two-year rebuild by the late Nick Grace, it flew again with the new registration G-BOML. Later that year it was purchased by the Old Flying Machine Company and has become a much sought-after airshow performer. The aircraft has also appeared in the movie *Memphis Belle* and on television in *Piece of Cake* and *Perfect Hero* and, more recently, in the French movie *The Diamond Swords*.

The Messerschmitt wore the colours of two Luftwaffe aces. The fuselage markings are those of Oberleutnant (later Major) Joachim Müncheberg, squadron commander of 7 Staffel (squadron) Jagdeschwader (fighter wing) 26, which operated over Malta in February 1941. At the time he was credited with

Above Lift-off and the culmination of many months of planning as, over 20,000 kilometres away from its home base, the Messerschmitt performs. Note the use of slats ahead of the wing leading edge. These provide extra lift on the outboard part of the wing when it is at a high angle of attack. As Mark Hanna explains, this allows good aileron control right up to the stall and obviates 'washout' (a twist) in wing, as in the Spitfire.

23 victories. On 23 March 1943 he died evading American fighters when his 109 shed its wings. His final tally stood at 135.

Müncheberg's Messerschmitts all featured a yellow-painted nose on which was superimposed a red heart (the squadron badge). They also carried the script S emblem of the Geschwader (the wing). The chevrons represent his rank and the bar aft of the Luftwaffe Cross, the fighter wing.

The rudder markings belong to Oberleutnant (later Major) Josef 'Sepp' Wurmheller who flew with Jagdeschwader 26 operating over the English Channel coast and then later over Russia. They depict an RAF roundel, a Russian star and the Knight's Cross of the Iron Cross with Oak Leaves. The inscription 1939 recognises that this award was instituted on 1 September 1939. The tally of aircraft shot down is 82 (60 plus 22). Wurmheller was killed in action — a mid-air collision — on 22 June 1944. His tally was 102.

The aircraft's undercarriage is ingeniously designed. The weight of the aircraft on the ground is carried by the fuselage, rather than by the wings as in other aircraft. The reason for this

Above Low and fast in the Hanna tradition. Note the square-framed canopy which allows very little headroom and limited visibility. The aircraft appears to be smiling.
Left G-BOML banks away from the camera. Its armament consisted of two 20-mm Hispano wing-mounted cannon with provision for eight 80-mm Oerlikon rockets.

design is that the wings were not strong enough to carry the aircraft's weight. The wings could be removed without other support, but the narrow undercarriage caused many landing and taxiing accidents.

At the conclusion of the airshow the Messerschmitt was flown back to Dunedin, dismantled once again by Roger Shepherd, then crated and shipped back to England to be made ready for its European Airshow commitments.

Left High drama as the Spitfire closes on the Messerschmitt in the skies above Wanaka.

Pacific Theatre Aircraft

Curtiss Warhawk/Kittyhawk

The AFC's Kittyhawk arrived in New Zealand in 1988 with only superficial restoration work having been done by its previous owner; the task to restore the old warbird completely was placed upon the shoulders of Alpine's engineers. Under the supervision of engineers Ray Mulqueen and Tony Ayers the team, which included Marnix Pyle (at the start), Wallace Harvey and Dave Conmee, began their task, little realising that it would take them four years and thousands of hours. Towards the end of the project Dave Collinson and Paul Jones joined the team. Projected completion dates were extended until Thursday 16 April 1992, when the aircraft was rolled out of its hangar and Ray Mulqueen flicked the V12 Allison into life for the first time in 50 years. This was a momentous occasion for the AFC. Simple problems were resolved and in the late afternoon of Friday, the unpainted aircraft (there just wasn't time), was rolled out of the hangar once again and engine and ground running tests were performed up to full power.

With tests completed, the midnight oil was burnt once again to make minor adjustments before the first flight on the morning of the airshow. (This is not the usual practice as it places extra strain on the nerves.) There was an expectant hush over the gathering crowd and last-minute anxiety from the engineers as Mark Hanna taxied out, undertook cockpit checks and then began the take-off run. As the wheels lifted off the runway, Alpine's engineers broke into excited applause with everyone congratulating themselves for a job well done. After two faultless test-flights the P-40 was wheeled back into the hangar for oil filter inspections etc., and was pronounced airworthy in time for its programmed display by Mark Hanna.

One of 200 of the K-5-CU model, this P-40, c/n 21117, USAAF No. 42-9733, was originally built in 1942 by Curtiss at their Buffalo plant and operated by the United States 11th Air Force. Flying over the Aleutian chain of islands, defending an area spanning the coastlines of Alaska and Russia, the aircraft ended its war crashing on Amchitka Island in the Aleutians.

Above Mark Hanna eases the P-40 into the air for its second flight.
Left Test flight debrief. Engineers and supporters group around P-40 test pilot Mark Hanna, following the aircraft's first flight. Standing on the left wing behind Mark Hanna are Wallace Harvey from the AFC and Paul Stridnell, an Australian working for Col Pay, owner of Australia's only airworthy P-40. Sitting on the starboard wing is Dave Collinson, AFC, while grouped, left to right, are Lisa Collinson, Ray Mulqueen (AFC), and Ray Hanna (OFMC).

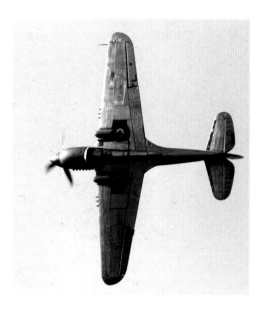

Above Symmetry in motion.

There it lay for many years until it was recovered in 1969 and shipped to the States. A rebuilding programme commenced in the early 1970s with the Kittyhawk, registered N4363, moving through a succession of owners — Bob Sturges, of Troutdale, Oregon; Joseph A. Morasky, Guildford, Connecticut; George Enhorning, Wolcott Air Services, Connecticut and later Byrne Aviation, Bloomfield Hills in Missouri. The AFC entered the picture in 1988, purchasing what they believed was a partly restored aircraft.

The aircraft wears the colours of NZ3108 18, a P-40K-1-CU operated by No. 16 Fighter Squadron. The original NZ3108, c/n 42-45959, began its service career with USAAF's 68th Pursuit Squadron, based on Tonga, in 1942, and was allotted the code number 25. The aircraft at this stage was painted in a combination of RAF colour scheme and USAAF markings and known as the Warhawk.

The New Zealand connection began in October 1942 when No. 15 Fighter Squadron RNZAF arrived on Tonga and inherited 25, plus the E and K models it served with. The Kittyhawks, as they became known, were part of a lend-lease agreement aimed at helping New Zealand to expand its fighter defences with the long-term goal of participating in the Pacific fighting alongside Australian and American forces.

In April 1943 the aircraft was acquired by No. 14 Fighter Squadron RNZAF operating off Pallikulo Strip (Bomber 1) on Espiritu Santo. In June 1943 the aircraft changed hands again, this time being operated by No. 16 Fighter Squadron RNZAF. By this time it had been repainted in RNZAF colours and carried the serial NZ3108 18.

No. 16 continued to fly from Espiritu Santo and it was here that 18's war ended on 23 June 1943 when its pilot, Sergeant Rowan A. Blair, was undertaking night flying practice. Pallikulo airstrip did not have a flarepath; the only assistance available to pilots was a dim light at the far end of the runway. Blair mistook the light of a parallel runway for the one on his and lined himself up for take-off. The aircraft predictably left the runway, crashing through a stormwater drain and continuing into the palm trees surrounding the airstrip. Sergeant Blair's safety harness was not tight enough and his faced smashed into the gunsight, almost removing his nose. The aircraft was a

write-off. Rowan Blair survived the war and died in 1985. (Special thanks to Warren Russell for the use of his research material.)

The upper surfaces of the Kittyhawk are painted olive drab with lower panels in neutral gray, the colours it wore at the time of the crash. The registration ZK-FRE was originally reserved for the Kittyhawk but was later changed to ZK-PXL. XL are the Roman numerals for 40, hence P-40.

ZK-PXL's aerial antics are not confined to the display line at Wanaka as over 500 North Island dairy farmers were to find out on 23 March 1993. The aircraft, which only minutes earlier had given an aerial display over Invercargill Airport, set course for a 'surprise' low-level attack on the Dacre farm of Paul and Barbara Crooks in the heart of Southland. The group had gathered for lunch and listened to guest speakers talk about the province's growing importance to the dairy industry. Tim's interest in this industry and his love of practical jokes led him

Above The Kittyhawk displaying its lines. The moment that makes four years of hard work worthwhile.

to involve the P-40 in some fun. Detonators and flash-bombs (usually seen at Wanaka during the airfield attack) were strategically placed around the area by Alpine's Peter Walthew. To counter the 'threat' from North Islanders moving south to buy up properties, the P-40, flown by John Lamont, 'attacked' Southland farmer Art Bloxham, who was using a toilet out in a field. With pants down around his ankles he quickly vacated the outhouse before it exploded. A hay stack also went up in flames. Shotgun in hand, Art returned fire on the North Island attack, 'shooting' down the aircraft, which exploded in a ball of flame.

Chance Vought Corsair

Known by its pilots as 'the bent-winged bastard from Connecticut', the 'ensign killer' and the 'widow maker' because of the large number of accidents in which it has been involved, the F4U Corsair did, however, earn respect from the Japanese, who referred to it as the 'whistling death'. The Corsair was arguably the best carrier-operated fighter of its time.

The aircraft was conceived and developed by the Vought engineering staff, under Rex B. Beisel, to US Navy specifications for a new high-performance single-seat shipboard fighter. The operational requirement for speed and altitude performance was exhibited when, on 1 October 1940, the prototype became the first US fighter to exceed 400 mph in level flight. The first production model was wheeled off the assembly line in June 1942.

The Alpine Fighter Collection's Corsair was built in 1943 as an F4U-1, USN Bureau number 17995, and has the very rare original 'birdcage canopy' (the canopy, when closed, gave pilots the feeling of being a caged bird) and is believed to be the only Corsair of this type flying today. The RNZAF did not operate any Corsairs with this type of canopy.

The prop's diameter of 13 feet 4 inches (Spitfire 10 feet 9 inches) led to the devising of the inverted gull-wing configuration to avoid the very long undercarriage legs that would otherwise have been necessary for the 2000 hp Pratt & Whitney R-2800-8 Double Wasp engine. The undercarriage is arranged

for the main legs to fold straight back, with the wheels rotating through 90° to lie flat in the wheel wells in the wing centre section.

The AFC's Corsair did not fire its six 20-mm machine guns in anger as it served in a training role during World War II, being based in the United States. At the end of the war the Corsair was struck off charge and presented to the town of Provo, Utah as a reward for the town's efforts in selling war bonds. The aircraft was eventually sold to a private purchaser, as the town could not agree on how it should be displayed, or how to meet the costs involved.

The aircraft eventually ended up derelict in the local junkyard until discovered in 1966 by Harry Doan, who purchased the remains. The long process of restoration to airworthy condition began. Some fifteen years and thousands of hours later, a beautifully restored Corsair took to the skies once again in 1981, registered as N90285 to Doan Helicopters of Daytona Beach, Florida. The rebuild incorporated the later model speci-

Above Tom Middleton lifts ZK-FUI into the air. The ground disappears at 90 knots and 44 inches boost.

fication undercarriage, stalling strake and tailstrut which, compared with the earlier model, make it easier to fly.

Ownership passed to Roy Stafford, of Jacksonville, Florida in 1989 and in July that year to D.K. Precision, of Abilene, Texas, and into the hangar of the late Don Knapp.

N90285 attracted the AFC's attention in 1990 and at the time of its acquisition in 1991, the aircraft had accumulated 850 hours of total time, 75 of which were flown since restoration.

The colour scheme chosen is that of NZ5201 1, the first F4U-1 operated by the RNZAF in the Pacific, flying with eleven of the thirteen RNZAF fighter squadrons. In its later days it wore the cowling No. 201 with the inscription 'Alma' on the forward port fuselage. The RNZAF was the only service to use the type exclusively in a land-based role with the tail arrestor-hook removed.

Surplus to requirements at the end of World War II, NZ5201 became one of 438 American-built former combat aircraft to be sold on 2 March 1948 in what David Duxbury, a well-known New Zealand aviation historian, referred to as 'the sale of the century'. They were acquired by H.J. Larsen, at Rukuhia, Hamilton, for £10,000 and broken up for scrap.

In early 1992 the Corsair flew for a Singapore beer company, appearing in an $800,000 television advertisement that involved an air race with the Mustang (ZK-TAF) and a Harvard (NZ1037), filmed in the skies above Wigram and Queenstown. Unlike the Mustang, the Corsair was not repainted, although a growling bulldog's head appeared in the centre of the fuselage roundels.

Above The Pratt & Whitney R-2800-8W. Eighteen cylinders of Double Wasp Radial, which the Japanese referred to as the 'whistling death'.

Top left The Corsair on short finals. At this stage forward visibility is almost non-existent. There is almost 3 metres of airframe between the pilot and the propeller.

Bottom left The aircraft is capable of 480 knots and +7 g, but will not be pushed beyond 320 knots or so and a +4 g, as its age dictates a certain respect.

Post-World War II Aircraft

Hawker 'Baghdad' Sea Fury

Built at Langley in 1952 for an Iraqi order and delivered to Iraq in December 1953 serialed 326, the Hawker 'Baghdad' Sea Fury c/n 37723 is considered by many to be the ultimate piston engined fighter.

Named the Sea Fury, which indicates the type's original connection to aircraft-carrier service, the aircraft had retained its

Above A thorough pre-flight check.

wing-folding and tail-hook mechanisms. It lacked the hydraulic operation, however, as it was built denavalised and operated with the Iraqi Air Force (hence Baghdad in the aircraft's name). No. 326 was withdrawn from service, going into storage after only 120 engine hours and there it stayed until acquired by Americans Ed Jurist and David Tallichet in 1979 and shipped to Florida. In 1984 the aircraft was registered N43SF to Vintage Aircraft International, Nyack, New York.

In June 1986 the aircraft made its way to New Zealand and into the ownership of Flightwatch Services Ltd. On 12 March 1988, following a full restoration at Ardmore, the Sea Fury, complete with original engine, made its maiden flight in the hands of the skilful Australian Guido Zuccoli. It is now a regular airshow performer, being displayed by part-owner Robbie

Booth. The aircraft wears the colours of the Royal Navy Sea Fury, WJ232 114 of HMS *Ocean* flown by Lieutenant Peter 'Hoagy' Carmichael RN, who shot down a MiG 15 jet fighter during the Korean War. An oil painting depicting the air battle by Robert Taylor is on display at the Fleet Air Arm Museum, Yeovilton, in Somerset, England.

In early 1992 the aircraft was fitted with the wing-folding mechanism from the remains of WG655, a two-seat Sea Fury Trainer. This aircraft flew with the Fleet Air Arm Historic Flight, and was written off in a forced landing in England on 14 July 1990.

Above The Sea Fury is eased off the runway by Robbie Booth.

103

de Havilland Australia Vampire

The Spider-Crab, as it was initially called, appeared on the de Havilland drawing boards in 1941. The prototype first flew in September 1943 but squadrons did not receive the aircraft until 1946, too late for war service.

The need for an advanced jet trainer was recognised and on 15 November 1950 the DH115 trainer prototype flew for the first time. In 1951 the DH100 entered service with the RNZAF, becoming the first operational jet fighter. Forty-seven DH100s and eleven DH115s served with numbers 14 and 75 squadrons until their retirement in 1972 before being sold overseas.

The appropriately named Vampire Syndicate was keen to redress this situation and in November 1990 purchased a DH-115 Vampire T35W from Australia (the W indicates self-sealing

Above High performance thrills. ZK-VAM is guided into the air by Ross Ewing and John Peterson. Sponsorship for the aircraft is a prerequisite as the jet can burn up to 1,000 litres of fuel an hour.
Left The Sea Fury was the Fleet Air Arm's last operational piston-powered fighter and arguably the ultimate.

105

Above The Vampire in flight. The aircraft was originally called the Spider-Crab.

fuel tanks). The aircraft is the first two-seat former military jet trainer to be privately owned in New Zealand and appears on the civil register as ZK-VAM.

The aircraft was flown across the Tasman by ex-RAAF pilot Bill Scott and Air New Zealand captain John Peterson, making the flight in two hops: Brisbane–Norfolk and Norfolk–Auckland. The flight was also a first for this type of aircraft, with touchdown at Auckland International Airport following a total flying time of less than four hours and into a customs inspection that was unusual for inspectors used to wide-bodied airliners.

The Vampire was built by de Havilland Australia at Bankstown in July 1959, and served with the RAAF as A79-649 for ten years, being operated by No. 25(F) Squadron at RAAF Pearce in Western Australia. It was withdrawn from service late in 1969 and sold in January 1970. Restored in 1986, it flew once again, joining the Australian civil register as VH-ICP and appearing at a number of airshows between late 1986 and early 1988, operated by Jecani Pty Ltd.

The Airfield Attack

Above Bill Rolfe flies his Harvard past a fireball.

The enemy, consisting of six Harvards and the Messerschmitt, attacked the airfield with carefully co-ordinated machine gun fire and bomb bursts. Then, to the delight of the crowd the good guys arrived on the scene: the Vampire, Corsair, Kittyhawk, Spitfire and Sea Fury joined the battle.

High-speed duels ensued and aircraft twisted and turned in the sky, taking evasive action in an attempt to shake off their attackers. At the height of the battle smoke rings from the explosions filled the sky and Mark Hanna in the Messerschmitt climbed, followed by his father Ray in the Spitfire. They pulled their aircraft up through a smoke ring, banked tightly and dived at high speed towards the ground. A Harvard, its engine streaming smoke, exploited the area's geography by 'crashing' into the Clutha River valley.

The enemy retreated.

Above Boom! The heat from the flash bombs could be felt 200 metres away.
Right 'And then the sky was empty...' The Messerschmitt has centre stage among explosions and smoke rings.

A Review

A very full day of flying and static displays was organised for Saturday, 18 April. The morning started for the pilots with a briefing at 9 o'clock. For the really keen enthusiasts the day began at 7 am (before sunrise) when the gates were opened. With the assistance of the young enthusiastic ATC cadets, car-parks were quickly found. Armed with cameras, souvenir pro-grammes and warm jackets, the spectators embarked on their tour.

A careful study of the airfield map in the highly informative programme revealed the location of the displays. Strolling around the airfield so early in the morning, the keen photogra-phers could capture on film those special shots, the ones where the pilot or engineer is working on their aircraft, with the fuse-lage panels open and the engine cowling removed.

On that particular Saturday there was an added incentive to arrive early: the sight of a Messerschmitt, Spitfire, Kittyhawk, Corsair and Sea Fury parked side by side and a Venom on the tarmac in front of the hangars. If told five years before that air-craft of this calibre would be assembled on an airfield in Cen-tral Otago, many would have been incredulous. But here they were, with an amazing nine Harvards and seven Tiger Moths thrown in; it was little wonder a crowd of 55,000 turned out. Sadly the Mustang did not arrive; it was grounded at Momona Airport, Dunedin.

Those who arrived later in the morning were greeted by long lines of traffic. The 2-km trip from Luggate, in the south, was taking an hour and vehicles were backed up as far as Wanaka in the north, 7 km away.

The trade displays included aircraft, engines and various assorted parts for sale. There were books, magazines, posters, postcards, pens, badges and T-shirts for the serious collectors. Food stalls kept the thousands from going hungry and a greater number of toilets were also available. Throughout the day the public address system played a variety of music with added information from a group of knowledgeable commenta-tors.

During the morning the senses were treated to the sights,

Above Trevor Bland, President, NZ Warbirds Association.

Above A classic formation probably never to be repeated — Corsair, Spitfire, Vampire, Messerschmitt and Sea Fury.

sounds and smells of aircraft only too eager to perform. The most memorable were Richard Hood and John Kelly, combining well in their two Pitts Specials, the successful world record attempt by Evan Bloomfield (general manager, A.J. Hackett Bungy, Cairns, Australia) and Ian Binnie (camera supervisor, Queenstown) to bungy jump from helicopters at 1,000 feet, the Kit Fox display given by 'Mr Kit Fox' himself, American Dan Denney, and the RNZAF's Kiwi Blue parachute display and the Red Checkers.

At midday the sun appeared, after heavy cloud, jackets were discarded and the water trucks arrived, spraying the taxiways to keep the dust away.

Mystery and intrigue surrounded the listed display at 12.55 pm, of a 'surprise overseas military fighter'— yet to be confirmed. Thirty minutes before its arrival time, Tim Wallis confirmed over the PA system that the aircraft was a Lockheed F-117 Stealth fighter, now in New Zealand airspace, having flown from Australia. The pilot was contacted and over the PA

Above Sixteen hectares of private aircraft parking.

system Major Conrad (alias Dan Denney) gave a commentary as he 'flew' down the North Island across Cook Strait and then proceeded to describe the Southern Alps. The major informed the crowd when he was within radar range of the airfield and scanning the area for his target — an old Valentine tank, parked on the far side of the runway. 'Locking onto' his target at five miles out he 'fired' a radar-guided missile and an explosion rocked the tank. Tim congratulated him on his accuracy and asked him if he would give a fly-past and land, but the major reported that he was low on fuel, gave his apologies and set course for Australia. The crowd came to realise that they had been had. Tim's humour had shone through.

The afternoon was devoted to the warbirds — the distinctive chop from the Harvards as their propeller blades went supersonic, the fine smooth whistle of the big-radialled Sea Fury, the curvaceous lines of the legendary Spitfire and the growl of the yellow-nosed Messerschmitt, the excited clapping by Alpine's engineers as their four-year restoration project — a Kittyhawk

Above A relaxed gathering of the main display teams, the Roaring Forties and Red Checkers. Sitting on the wing left to right, John Lamont, Flight Lieutenant Phil Murray and Keith Skilling. Standing left to right, Steve Taylor, Flight Lieutenant Peter Walters, Squadron Leader Hoani Hipango, Flight Lieutenant Nick Osborne, John Peterson, Squadron Leader Ian McClelland, Robbie Booth and Flight Lieutenant Mike Tomoana.

— leapt into the sky for the first time in nearly 50 years, the gull-winged Corsair and the scream of the fork-tailed Vampire.

But the display that had the audience's adrenalin pumping the hardest was the dogfight between the Spitfire and the Messerschmitt. These two aircraft became more than just stories in war books and comics, as their skilful pilots recreated the realism of the tense and deadly air battles that took place over Europe during World War II. During the display, with the songs of Dame Vera Lynn playing in the background, two former New Zealand fighter pilots wept.

Sunday was a more relaxed day, with most of the aircraft performing again. The day finished with 'Reno style' air races. The feature race was won by the Sea Fury, even if it did cut a few corners!

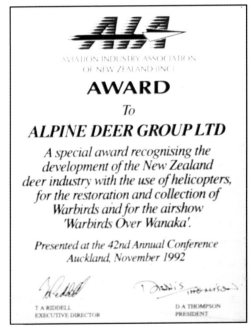

Top left The crowd, with eager photographer focusing on the 109J.
Top right Civic Development Award.
Above Aviation Industry Association of New Zealand (Inc.) Award, 1992.
Left New Zealand Tourism Award.

113

New Zealand Fighter Pilots Museum Opening and Fly-in 1993

The air is still. The wind sock hangs limp. The sun's warmth begins to break the coolness of the morning. The only sounds to be heard are birds in the distant trees. The moon lies low in the pale blue sky. One by one the Alpine team arrive at the airfield to complete last minute preparations for the busy day ahead.

Saturday, 10 April, 11.40 am and the thunderous roar of a Merlin shatters the silence as the Mustang, making a welcome return to Wanaka, 'bounces' the airfield, swooping in low from behind the hangars. Climbing high into the sky, the aircraft performs a victory roll, ending the three-year interval since its last performance above the airfield.

At the controls was Trevor Bland, President of the New Zealand Warbirds Association; sharing the experience in the back seat was part-owner, Air New Zealand 747 captain John Sager. Cruising at 18,000 feet, the flight from Ardmore took them 1 hour 55 minutes.

That Easter Weekend 1993 saw the official opening of the New Zealand Fighter Pilots Museum. Three hundred guests, including many former fighter pilots and their wives, attended the private function held on the Saturday.

Tim Wallis opened the ceremony by outlining the aims of the museum: 'To collect, preserve and display information and memorabilia relating to those New Zealanders who flew fighter aircraft in our defence during the First and Second World Wars. The museum honours our warriors. It does not seek to glorify war.' He added, 'The museum has been established to complement the Alpine Fighter Collection of classic warbird aircraft.

Right The P-40 Kittyhawk, with Wanaka hills in the background.

New Zealand Fighter Pilots Museum
A P-40 Kittyhawk against a backdrop inspired by the New Zealand flag. The Kittyhawk was chosen because most New Zealand fighter pilots flew this type of aircraft in the Pacific theatre of World War II.

People would turn up and ask to see the Spitfire they had heard was based at the airfield. One or two people was no problem but when numbers increased to over twenty a day we couldn't get any work done.' This led to the appointment of Ian Brodie as manager of the New Zealand Fighter Pilots Museum. Ian, a member of the Warbirds Association, has been photographing their aircraft for many years. Employed by Air New Zealand for fourteen years in Christchurch, he ended his career there as electronic distribution systems co-ordinator, New Zealand and Pacific Islands region.

Tim also spoke highly of the dedication of Nola Sims, his secretary for 25 years, and the skill of Ray Mulqueen, his multi-talented chief engineer. Ray has 25 years' experience in aircraft engineering, and some 1,400 flying hours. He joined the Alpine Deer Group in November 1988, following a two and a half year period in Australia assisting Warbird enthusiast Col Pay with the restoration of his P-40. His appointment at Wanaka was timed to coincide with the arrival of the Spitfire from England.

On Easter Sunday a number of the AFC's pilots flew the aircraft to keep their ratings current, providing a relaxed display.

CCF Harvard IV/A6M2 'Zeke'

The Mitsubishi Type 0 'Zeke' owes its name to a unique numbering system adopted by the Japanese. Aircraft numbers were allotted based on AD 1925 which is the Japanese year 2585. AD 1940 (the year the aircraft entered full service with the Imperial Japanese Navy Air Force) is the Japanese 2600 and the aircraft has taken the last figure 0 as its type number. The Allies gave the aircraft the codename 'Zeke'. It was known as the Type Zero Fighter (Rei Shiki Sento Ki) which was often abbreviated to Rei-sen and commonly called the 'Zero-sen'.

The AFC's aircraft began its life as a Harvard IV, N15798, in the early 1950s on the production line at the Canadian Car and Foundry Company, c/n CCF4-1453, and served with the RCAF, as 20362.

The Harvard has a number of interesting modifications made so that it would resemble a Mitsubishi A6M2 for the 1970 film *Tora! Tora! Tora!* Its cosmetic conversions include a geared

Pratt & Whitney R-1340-61 engine, a three-bladed prop, machine guns on the forward fuselage above the engine, longer wings and a retractable tail wheel, giving the aircraft an additional 30 knots over the standard T.6 Harvard. The second cockpit still remains, very cleverly hidden under artificial armour and rear fuselage decking. The Zeke has a bomb rack and bomb.

A6M2 stands for (A) carrier-borne fighter, (6) sixth to go into service with the Imperial Japanese Navy Air Force, (M) manufactured by Mitsubishi, (2) second model of this design. It wears the authentic dark green colour scheme with the tail code 288, as operated from Rabaul in 1943. The 'enemy' aircraft has also appeared in the films *Midway* and *Baa Baa Black Sheep* and on television in the mini series *Winds of War* and *War and Remembrance*.

At 550 hp N15798 is considered underpowered (the original A6M Zeke produced 1130 hp) for realistic dogfights with the P-40 (1325 hp) and there is the possibility of re-engining it with an ex-DC 3 1200 hp P & W R1830. Its machine guns will also come 'alive' with the assistance of butane gas guns in the wings, giving a convincing sound effect. There is potential for some exciting dogfights with the P-40.

Above left Ian Brodie preparing a static display on fighter planes for the museum.

Above right Ray Mulqueen delving into the cockpit of one of the planes on display.

Grumman Avenger

Known as the 'Turkey' to its wartime pilots, the Grumman Avenger was, however, known for its ruggedness as it frequently continued flying after withstanding heavy battle damage.

Built in 1945 by General Motors Eastern Aircraft Division, Trenton New Jersey, Bu. No. 91110, the Avenger served in the US Navy with VA-22 for many years. In 1963, as N6827C, it was acquired by TBM Inc. of Tulare, California, a large operator of the type for agricultural and fire-fighting use. In 1977 it went to Hillcrest Aircraft Co., Lewiston, Idaho, and in 1986 to Paramount Leasing Corp of Bakerfield and Gro Pro Corp, Oklahoma City. The aircraft was acquired by the OFMC in May 1988 and was initially painted in spurious Royal Navy markings V110/BA.

ZK-TBM has a complete documented history, with all the original log books available. When bought by the AFC it carried the markings of one of the US Navy's more celebrated fliers, a future President of the United States, Lieutenant (J.G.) George Bush.

Acquired by the AFC from the OFMC in 1992, the aircraft, apart from new hoses and pipes, is in original condition, with gun turret, 0.5 inch machine gun, and operational bomb doors.

The Avenger was repainted in April 1993 in the colour scheme of Flight Lieutenant Fred Ladd's 'Plonky', NZ2518, which he flew in 1944 while serving with 30 Squadron in the Pacific. As Fred recalled in *A Shower of Spray and We're Away*:

> I fell in love at first sight with my machine, No. 2518, although it was undoubtedly the rustiest and oldest looking TBF in the whole squadron. Everyone remarked upon its scruffiness and everybody who flew it said it felt 'rusty', but to me it was a wonderful machine.
>
> Soon after each crew had been allocated its kite it got busy providing it with suitable insignia. We gave ours a cartoon-type drawing of an animated beer barrel, complete with wings, warrior-like face, and a spout for a nose. The spout was pouring grog at a great rate and we called our kite 'Plonky'. In those days I didn't drink alcoholic beverages and the insignia por-

Left Simon Spencer-Bower in N15798, complete with bomb, makes a low pass. The registration ZK-ZRO has been reserved for the aircraft.

Right John Lamont folds the Corsair's wings. These allowed the aircraft to be stored in the cramped hangars aboard aircraft carriers.

Below Tom Middleton folds the wings of the Avenger. The Wright R-2600-20 engine pumps out the oil.

trayed 'Plonky' pouring stale beer all over the Japs — about the most insulting thing I could think up for her to do to them . . .

The crew's last operational sortie together, their thirty-second, was on 22 May 1944, over Talili Bay supply area. No. 30 Squadron returned to New Zealand, leaving their Avengers to an American squadron who flew 22 sorties in the aircraft on 22–24 May over Bougainville and Rabaul before handing them over to No. 31 Squadron.

Fred Ladd wrote: 'I was rather sad to leave "Plonky" behind and I think she may have shed an oily tear or two at seeing us go. We'd been attached for what seemed a pretty long time. Shortly after we left, she was shot down over Rabaul with her new crew. One of the wings was seen to come off, she ploughed in and all aboard were killed.'

The AFC's TBM-3E differs slightly from the 48 TBF 1 and 3's operated by the RNZAF. The most recognisable difference is the redesigned cowling to accommodate the repositioned oil cooler. The oil cooler intake was moved to the bottom lip of the cowling. The TBMs were manufactured by the General Motors Eastern Aircraft Division and the TBFs by Grumman.

'Plonky' has the call-sign 'Avenger One-Eight' and is recognised by its deep chest, big glasshouse, square-cut surfaces and high tail.

Warbirds Over Wanaka 1994 — A Preview

Yakovlev Yak-3

The Russian Yakovlev Yak-3U is an interesting recent addition to the AFC, and the smallest and lightest fighter to see large-scale operational service during World War II.

The AFC aircraft is not a restored fighter but one actually built up from newly fabricated components in the original wartime factory at Orenburg in the Urals. One concession to the otherwise faithful duplicate is the power plant chosen. Klimov engines are now extremely rare so an Allison V-1710-99, noted for its reliability, has been fitted. Some modified instrumentation for New Zealand requirements has been installed in the otherwise authentic Russian cockpit.

Test-flying of the prototype began in Russia in early 1993. The AFC's aircraft is the second off the production line.

Cessna A-37B Dragonfly

Manufactured 28 August 1972, the Dragonfly, serialled 71-854, is a two-seat trainer that served in Vietnam with US 604th Air Commando Squadron at Bien Hoa. It was later allocated to the South Vietnamese Air Force, and was then captured. The Cessna, registered ZK-JTL, now owned by Jet Trainers Ltd, came from Vietnam with 1108 hours on its airframe, and a small selection of spares, including two engines.

Mikoyan-Gurevich MiG 15

The MiG-15 UTI has been acquired by Mike Kelly and Murray Patterson of Southair Aviation. It is an ex-Polish two-seat trainer that has been refurbished and reassembled by Winrye Aviation, Bankstown Airport, Sydney, Australia. The aircraft, registered ZK-MIG, is to be based at Wanaka.

Appendices

The appendices list the aircraft featured in the flying displays and a small number of interesting statics. Some aircraft, although parked within the display enclosure, were not part of the main display so have not been listed, for example a small number of Cessnas.

The aircraft are listed alphabetically under manufacturers' names, followed by their registration numbers, the display pilot(s) over the weekend, the aircraft's owner and base at the time of the airshow.

A note about registration and serial numbers carried by the aircraft listed in the appendices:

ZM was a special allotment for the experimental
  Flying Flea aircraft.
ZK indicates New Zealand civil register.
NZ indicates New Zealand military serial.
VH indicates Australian civil register.
A indicates Australian military serial.
G indicates Great Britain permanent civil register.
K, WB etc indicate Great Britain military serial.
N indicates United States civil register.

Where two registrations are listed, for example for the Harvards, this is because the former military aircraft now appear on the New Zealand civil register.

The appendices have been compiled from a large number of sources and every effort has been made to ensure their complete accuracy. I apologise to any aircraft owner or pilot who may have been incorrectly listed or omitted from the lists. Please let me know, so the correction can be made in later editions.

The more serious aircraft observer will note that some aircraft listings in the souvenir programmes differ slightly from those that appeared at the airshows. These have been corrected in this book.

Above Bill Black, one of New Zealand's most experienced pilots, with 21,000 hours on helicopters and over 3,000 hours fixed wing, has displayed at the three airshows.

Right CT-4B Airtrainer soloist Red 5 'Tux' Tomoana performing a vertical 360° and stall turn during the 1992 airshow.

Appendix 1
Warbirds On Parade 1988

Aircraft	Registration	Pilot(s)	Owner(s) and Base
Aero Engine Service Ltd AESL CT-4 Airtrainer	ZK-DGY	B.J. Rhodes	CT-4 Syndicate, Ardmore
Auster J.1 Autocrat	ZK-BJL	Tom Middleton	Tom Middleton, Temuka
British Aerospace BAe 125-800B	ZK-TCB	Barry Newsham	Wilson Neill Ltd, Auckland
Britten Norman BN2A-26 Islander	ZK-EVO	Ross Taylor	Aspiring Air 1981 Ltd, Wanaka
Cessna U206G Stationair	ZK-EFI	Ken Calder	Waterwings, Te Anau
Cessna U206G Stationair	ZK-KPM	Russell Baker	Air Fiordland, Te Anau
de Havilland DH82A Tiger Moth	ZK-BLI	Sir Peter Elworthy	Sir Peter Elworthy, South Canterbury
de Havilland DH82A Tiger Moth	ZK-BRB(NZ1459)	Tom Middleton	Alpine Helicopters Ltd, Queenstown
de Havilland DH82A Tiger Moth	ZK-BRC	Jules Tapper	Mount Cook Group, Queenstown
de Havilland DH82A Tiger Moth	ZK-CDU	Tony Renouf & Cheryl ('Rusty') Butterworth (wing-walker)	Tony Renouf, Ardmore
de Havilland DH89A/B Dragon Rapide/Dominie 'Tui'	ZK-AKY	Tom Williams	Tom Williams, Masterton
de Havilland DH112 Venom Mk FB.1	ZK-VNM(WE434)	John Denton	Trevor Bland, Ardmore
de Havilland Canada DHC-1 Chipmunk Mk.22	ZK-BSV	Eddie Cannon-James	Eddie Cannon-James, Auckland
de Havilland Canada DHC-1 Chipmunk T.10	ZK-DUC	Simon Spencer-Bower	Simon Spencer-Bower, Rangiora
de Havilland Canada DHC-2 Beaver	ZK-CKH(NZ6001)	Peter Rhodes	Warbirds Syndicate, Ardmore
Douglas C-47/DC-3 Dakota	ZK-DAK(26840)	Bill Howell, Keith Walsh, Paul Radley	Warbirds DC-3 Syndicate, Ardmore
Fletcher FU-24/950M	ZK-BHK	Bryan Beck	Central Airspread, Alexandra
Fletcher FU-24/950	ZK-DMO	Phil Brown	Central Airspread, Alexandra
Fokker F27-500F Friendship 'Korimako'	ZK-NFI	Captain Dave Offwood & First Officer Brian Horrell	Air New Zealand

Aircraft	Registration	Pilot(s)	Owner(s) and Base
Glaser-Dirks DG-400 (motorised glider)	ZK-GOM	Bruce Drake	Murray Wills, Invercargill
Hawker 'Baghdad' Sea Fury Mk FB.11	ZK-SFR(WJ232)	Robbie Booth	Fury Syndicate, Ardmore
Isaacs Fury II (7/10 scale replica)	ZK-JHR(K2059)	Rex Carswell	Hawker Syndicate, Hobsonville
Isaacs Fury II	ZK-RFC(K2046)	Dave Simpson	Dave Simpson, North Shore Airfield
Jodel D.9 Bebe	ZK-AKR	Charlie Kenny	Charlie Kenny, Clydevale, South Otago
North American AT-6D Harvard Mk III	ZK-ENE(NZ1066)	Keith Skilling	Booth/Porter/Rolfe Syndicate, Ardmore
North American AT-6D Harvard Mk III	ZK-ENF(NZ1065)	John Lamont	Broadbent/Duncan Syndicate, Ardmore
North American AT-6D Harvard Mk III	ZK-ENG(NZ1078)	Steve Taylor	Harvard 78 Syndicate, Ardmore
North American AT-6D Harvard Mk III	ZK-ENJ(NZ1098)	John Denton	K.R. Brooking, Ardmore
North American AT-6C Harvard Mk IIA	ZK-ENN(NZ1025)	John Greenstreet	John Greenstreet, Ardmore
North American AT-6D Harvard Mk III	ZK-WAR(NZ1092)	John Peterson	Bland/Schroder Syndicate, Ardmore
North American P-51D Mustang	ZK-TAF(NZ2415)	Trevor Bland	NZ Historic Aircraft Trust, Ardmore
Pilatus PC6/B2-H2 Turbo Porter	ZK-MCT	Phil Cooney	Mount Cook Group, Queenstown
Piper PA-18-95 Super Cub	ZK-BTM	Phil Murray	Southern Districts Aero Club, Gore
Replica Plans S.E.5a (7/10 scale replica)	ZK-TOM	Tom Grant	Tom Grant, Taieri
Thorp T-18 Tiger	ZK-KID	Stuart Kidd	Stuart Kidd, Invercargill

Interesting Statics

Aircraft	Registration	Pilot(s)	Owner(s) and Base
Spencer Amphibian Air Car Model 2-12-E	ZK-NMF		Neil Falconer, Gore
Beagle A109 Airedale	ZK-CCW		C.G. Mills, Gore

Helicopters

Aerospatiale AS350B Ecureuil (Squirrel)	ZK-HMY	Bill Black	The Helicopter Line, Te Anau
Hiller UH-12E 'Herbie'	ZK-HBL	Tim Wallis	Alpine Deer Group, Wanaka
Hughes 369D Model 500D	ZK-HOT	Tim Wallis	Alpine Deer Group, Wanaka
Robinson R22B Beta	ZK-HZT	Doug Maxwell	ACE International Partnership, Alexandra

Royal New Zealand Air Force

Aero Engine Service Ltd AESL CT-4B Airtrainers
Central Flying School and Pilot Training Squadron, RNZAF Base Wigram

'Red Checkers'	Red 1	NZ1943	Sqn.Ldr. Roger Read 'RR'
	Red 2	NZ1944	Sqn.Ldr. Michael Panther 'Merp'
	Red 3	NZ1933	Flt.Lt. John Benfell 'Benny'
	Red 4	NZ1936	Flt.Lt. Mark Wingrove 'Wing'
	Red 5	NZ1934	Flt.Lt. Ian Walls 'Wally'
	Red 6	NZ1931	Sqn.Ldr. Hamish Brice 'Hamie'

Bell UH-1H Iroquois
No. 3 Squadron Detachment, RNZAF Base Wigram

NZ3803	Sqn.Ldr. Paul C. Martin, Flt.Lt. Brian L. Coulter, F/SHCM Merv K. Baker

British Aircraft Corporation BAC 167 Strikemaster MK88
No. 14 Squadron, RNZAF Base Ohakea

NZ6363	Plt.Off. Stephen Alderton (static)
NZ6374	Flt.Lt. Dave Brown (display pilot)

Fokker F27-100 Friendship
Navigation Air Electronics and Telecommunications Training Squadron, RNZAF Base Wigram

NZ2783	Wg.Cdr. Noel McKenzie, Flt.Lt. Ian MacGregor, Flt.Lt. Dave Ashton (Nav.)

Appendix 2
Fly-in 1989

Aircraft	Registration	Pilot(s)	Owner(s) and Base
de Havilland DH82A Tiger Moth	ZK-BRB(NZ1459)	Tim Wallis	Alpine Deer Group, Wanaka
de Havilland DH104 Devon NT.1	ZK-KTT(NZ1808)	Barry Keay	Dalziell/Brown Syndicate, Ardmore
de Havilland Canada DHC-2 Beaver	ZK-CKH(NZ6001)	Peter Rhodes	Warbirds Syndicate, Ardmore
North American AT-6D Harvard Mk III	ZK-ENF(NZ1065)	John Lamont	Broadbent/Duncan Syndicate, Ardmore
North American AT-6D Harvard Mk III	ZK-ENG(NZ1078)	Steve Taylor	Harvard 78 Syndicate, Ardmore
North American AT-6D Harvard Mk III	ZK-ENJ(NZ1098)	Keith Skilling	K.R. Brooking, Ardmore
North American AT-6C Harvard Mk IIA	ZK-ENN(NZ1025)	John Greenstreet	John Greenstreet, Ardmore
North American AT-6D Harvard Mk III	ZK-WAR(NZ1092)	John Peterson	Bland/Schroder Syndicate, Ardmore
North American P-51D Mustang	ZK-TAF(NZ2415)	Trevor Bland	New Zealand Historic Aircraft Trust, Ardmore

Helicopters

Hughes 369D Model 500D	ZK-HOT	Tim Wallis	Alpine Deer Group, Wanaka

Royal New Zealand Air Force 'Wise Owl 48'
Aero Engine Service Ltd AESL CT-4B Airtrainer
Central Flying School and Pilot Training Squadron, RNZAF Base Wigram

NZ1931	Flt.Lt. Ricky Smits PTS
NZ1932	Flt.Lt. Chris O'Brien PTS
NZ1934	Flt.Lt. Keith Adair CFS
NZ1936	Flt.Lt. Dennis Butler PTS
NZ1939	Flt.Lt. Steve Morrissey PTS
NZ1941	Flt.Lt. Dave Barham PTS
NZ1942	Flt.Lt. Mark Woodhouse PTS
NZ1943	Flt.Lt. Warren Harre PTS
NZ1946	Flt.Lt. John Benfell PTS

Royal New Zealand Air Force — continued

NZ1947	Flt.Lt. Ron Thacker CFS	
NZ1948	Sqn.Ldr. John McWilliam OC PTS	

Bell 47G Sioux Helicopter
RNZAF Flying Training Wing, RNZAF Base Wigram

NZ3702	Flt.Lt. Keith Adair
NZ3706	Flt.Lt. Ron Thacker

Bell UH-1H Iroquois
No. 3 Squadron Detachment, RNZAF Base Wigram

NZ3808	Flg.Off. Phil Murray, Flg.Off. Mark Cook, M/HCM Chris Brownie

Fokker F27-100 Friendship
Navigation Air Electronics and Telecommunications Training Squadron, RNZAF Base Wigram

NZ2781	Flt.Lt. Terry Paddy, Plt.Off. Wayne Wallis, Flg.Off. Vaughan Davis, Flt.Lt. Murray Wratt, Plt.Off. Alistair Reilly
NZ2782	Flt.Lt. Pat McLuskie, Flg.Off. Julian Reynolds, Plt.Off. Grant Reedy

Appendix 3
Warbirds Over Wanaka 1990

Aircraft	Registration	Pilot(s)	Owner(s) and Base
Aero Engine Service Ltd AESL CT-4 Airtrainer	ZK-DGY	Peter Houghton	CT-4 Syndicate, Ardmore
Britten Norman BN2A-26 Islander	ZK-EVO	Alistair McMillan	Aspiring Air 1981 Ltd, Wanaka
Cessna 180A	ZK-BUJ	Dick Davidson	Dick Davidson, Moa Flat, Otago
Cessna 180C	ZK-BYI	Ernie Colling	Ernie Colling, Cromwell
Cessna A185F Skywagon II	ZK-DOC	Russell Baker	Bullmore & Wilkins, Balfour, Southland

Aircraft	Registration	Pilot(s)	Owner(s) and Base
Cessna A185E Skywagon II	ZK-DYH	Hank Sproull	Hollyford Tourist Company, Te Anau
Cessna A185F Ag-Carryall	ZK-JKH	Peter Bowmar	Peter Bowmar, Waikaia
Cessna A185F Skywagon II	ZK-MCU	Wayne McMillan	Mount Cook Group, Te Anau
Cessna U206F Stationair	ZK-DFW	Ken Calder	Waterwings, Queenstown
Cessna U206G Stationair	ZK-ETN	Brian Hore	B.L. & M.A. Hore, Nokomai Station, Lumsden
Cessna U206G Stationair	ZK-KPM	Russell Baker	Air Fiordland, Queenstown
Dassault Falcon Mystere 10	VH-MEI	Barry Newsham	Wilson Neill Ltd, Auckland
de Havilland DH60M Moth	ZK-AEJ	Bill Shaw	R. Gerald Grocott, Mandeville
de Havilland DH82A Tiger Moth	ZK-AKC	John Dixon	Allied Press Ltd, Taieri
de Havilland DH82A Tiger Moth	ZK-BAH	John Baynes	John Baynes, Otama, Southland
de Havilland DH82A Tiger Moth	ZK-BGP	Phil Keane	J.J. Mcleay Trust, Mandeville
de Havilland DH82A Tiger Moth	ZK-BLI(NZ1448)	Sir Peter Elworthy	Sir Peter Elworthy, South Canterbury
de Havilland DH82A Tiger Moth	ZK-BRB(NZ1459)	Andy Woods	Alpine Deer Group, Wanaka
de Havilland DH82A Tiger Moth	ZK-CDU	Tony Renouf & Cheryl ('Rusty') Butterworth (wing-walker)	Tony Renouf, Ardmore
de Havilland DH89A/B Dragon Rapide/Dominie 'Tui'	ZK-AKY	Neil Robertson	Tom Williams, leased to Croydon Aircraft Company, Mandeville
de Havilland DH104 Devon NT.1	ZK-KTT(NZ1808)	Peter McVinnie	Dalziell/Brown Syndicate, Ardmore
de Havilland DH112 Mk FB.1 Venom	ZK-VNM(WE434)	John Denton	Trevor Bland, Auckland
de Havilland Canada DHC-1 Chipmunk T.10	ZK-DUC	Simon Spencer-Bower	Simon Spencer-Bower, Claxby, Rangiora
de Havilland Canada DHC-2 Beaver	ZK-CKH(NZ6001)	Peter Rhodes	NZ Warbirds Syndicate, Auckland
Douglas C-47/DC-3 Dakota	ZK-DAK(26840)	Bill Howell, Jim Pavitt	Flightline Aviation, Ardmore
Fletcher FU24A-950M	ZK-CLO	Bryan Beck	Central Airspread, Alexandra
Fletcher FU24-950M	ZK-DMO	Ray Parsons	Central Airspread, Alexandra
Fletcher FU24-950M	ZK-DZF	Bob Clelland	South Otago Aerial Topdressing Co-operative Ltd, Balclutha
Glaser-Dirks DG400 (motorised glider)	ZK-GOM	Bruce Drake	Murray Wills, Invercargill

Aircraft	Registration	Pilot(s)	Owner(s) and Base
Government Aircraft Factory GAF N24A-60 Nomad	ZK-NMD	Tom Middleton	Air Safaris and Services (NZ) Ltd, Lake Tekapo
Hawker 'Baghdad' Sea Fury Mk FB.11	ZK-SFR(WJ232)	Robbie Booth	Fury Syndicate, Auckland
Isaacs Fury II (7/10 scale replica)	ZK-JHR(K2059)	Rex Carswell	Hawker Syndicate, Hobsonville
Jodel D.9 Bebe	ZK-AKR	Charlie Kenny	Charlie Kenny, Clydevale, South Otago
New Zealand Aerospace Industries NZAI Cresco 08-600	ZK-LTQ	Malcolm Hill	Turbo Air Services, Alexandra
North American AT-6D Harvard Mk III	ZK-ENG(NZ1078)	Steve Taylor	Harvard 78 Syndicate, Ardmore
North American AT-6D Harvard Mk III	ZK-ENF(NZ1065)	John Lamont	Broadbent/Duncan Syndicate, Ardmore
North American AT-6D Harvard Mk III	ZK-ENJ(NZ1098)	John Denton	K.R. Brooking, Ardmore
North American AT-6D Harvard Mk III	ZK-ENK(NZ1099)	John Peterson	Charles Darby and Jim Pavitt, Ardmore
North American AT-6D Harvard Mk III	ZK-WAR(NZ1092)	Keith Skilling	Bland/Schroder Syndicate, Ardmore
North American P-51D Mustang	ZK-TAF(NZ2415)	Trevor Bland	New Zealand Historic Aircraft Trust, Ardmore
North American T-28C Trojan	ZK-JGS(140563)	Keith Skilling	Estate of John Greenstreet, Ardmore
Piel CP-301 Emeraude	ZK-CBP	Jack Mehlhopt	Jack Mehlhopt, Timaru
Pilatus PC6/B2-H2 Turbo Porter	ZK-MCT	Phil Cooney	The Mount Cook Group, Queenstown
Piper PA-18A-150 Super Cub	ZK-BRX	Phil Murray	Ross Craig, Greenvale, Otago
Piper PA-46 Malibu	N2482Y	Brent Ferguson	Brent Ferguson Motors, Nelson
Pitts Special S1	ZK-EES	Richard Hood	R. & M. Hood and J. Kelly, Ardmore
Redfern Fokker Fok. DR.1 (full-scale replica)	ZK-FOK	Colin Glasgow	ZK-FOK Syndicate, Ardmore
Replica Plans S.E.5a (7/10 scale replica)	ZK-TOM(B168)	Tom Grant	Tom Grant, Dunedin
Stodart-Hamilton SH-2 Glasair	ZK-ADY	Tom Middleton	Mark Elworthy, Timaru
Thorp T-18 Tiger	ZK-KID	Stuart Kidd	Stuart Kidd, Invercargill
Vickers Supermarine Spitfire Mk LF XVIe	ZK-XVI(TB863)	Stephen Grey	Alpine Deer Group, Wanaka

Parachutists
(in jump order) John Britten, Brent Findlay, Dave Davies, Rob Noble, Julie Smith, Graeme Bates, Alastair Kay, Mike Humphrey, Koen De Smit, Barry Thompson, Brett Kennedy (drop zone controller); of the Otago Skydiving Club, Southland Parachute Club, and Christchurch Parachute Club

Interesting Statics

Aerotek Pitts Special S-1S	ZK-FRJ	Ashburton Aerobatic Syndicate, Ashburton
Beagle A109 Airedale	ZK-CCW	C.G. Mills, Gore
Curtiss P-40K-5-CU Kittyhawk	ZK-PXL	Alpine Deer Group, Wanaka (under restoration)
de Havilland DH104 Devon C1	ZK-UDO	Devon Syndicate, Ardmore
Spencer Amphibian Air Car Model 2-12-E	ZK-NMF	Neil Falconer, Gore

Helicopters

Aerospatiale AS350B Ecureuil (Squirrel)	ZK-HMY	Bill Black	The Helicopter Line, Te Anau
Hiller UH-12E 'Herbie'	ZK-HBL	Tim Wallis	Alpine Deer Group, Wanaka
Hughes 269C Model 300C	ZK-HQJ	Bryan Beck	Dunstan Peaks Ltd, Omarama
Hughes 369D Model 500D	ZK-HOT	Tim Wallis	Alpine Deer Group, Wanaka
Hughes 369D Model 500D	ZK-HLP	Dennis Egerton	The Helicopter Line, Queenstown
Robinson R22B Beta	ZK-HKM	Doug Maxwell	C.D. & N.J. Wyn, Haast

Royal New Zealand Air Force
Aero Engine Service Ltd AESL Airtourer T.6/24
Central Flying School, RNZAF Base Wigram

NZ1762	Static

Aero Engine Service Ltd AESL CT-4B Airtrainer
Central Flying School and Pilot Training Squadron, RNZAF Base Wigram

NZ1942	Flt.Lt. Peter Mount PTS

Bell UH-1H Iroquois
No. 3 Squadron Detachment, RNZAF Base Wigram

NZ3814	Flt.Lt. Dave R. Horrell, Plt.Off. Paul L. Stockly, M/HCM John W. Cooper, Plt.Off. Angie M. Dickinson

Royal New Zealand Air Force — continued
British Aircraft Corporation BAC 167 Strikemaster Mk 88
No. 14 Squadron, RNZAF Base Ohakea

NZ6372	Flt.Lt. Graeme Perry (display pilot)
NZ6374	Plt.Off. Ian Munro (static)

Fokker F27-100 Friendship
Navigation Air Electronics and Telecommunications Training Squadron, RNZAF Base Wigram

NZ2783	Flt.Lt. Pat McLuskie, Flg.Off. Brett Marshall

McDonnell Douglas A-4K Skyhawk
No. 2 Squadron, RNZAF Base Ohakea

NZ6205	Flt.Lt. Robert V. Jackson (tanker pilot)
NZ6217	Sqdn.Ldr. Ian Gore (display pilot)

Appendix 4
Warbirds Over Wanaka 1992

Aircraft	Registration	Pilot(s)	Owner(s) and Base
Altavia PZL 104 Wilga 35A	ZK-PZL	Dougal Dallison	Alternate Aviation, Taupo
Britten Norman BN2A-26 Islander	ZK-EVO	Alastair McMillan	Aspiring Air, Wanaka
Cessna L-19A/O-1G Model 305A Bird Dog	ZK-FYA(116903)	John Denton	Bird Dog Aviation, Ardmore
Cessna 180A	ZK-BUG	Hunter McEwan	Hunter McEwan, Auckland
Cessna 180H	ZK-PKD	George Galpin	George Galpin, Marton
Cessna A185E Skywagon	ZK-CYA	Hank Sproull	Willie J. Grey, Mossburn
Chance Vought F4U-1 Corsair	ZK-FUI(NZ5201)	Tom Middleton, Keith Skilling	Alpine Fighter Collection, Wanaka
Christen Industries Pitts S-2B	ZK-MAD	Richard Hood	The Great Stunt Company, Ardmore
Curtiss P-40K-5-CU Kittyhawk	ZK-PXL(NZ3108)	Mark Hanna	Alpine Fighter Collection, Wanaka

Aircraft	Registration	Pilot(s)	Owner(s) and Base
de Havilland DH60M Moth	ZK-AEJ	Murray 'Mo' Gardner	R. Gerald Grocott, Mandeville
de Havilland DH82A Tiger Moth	ZK-ALK	Rex Dovey	Murray Gardner, Queenstown
de Havilland DH82A Tiger Moth	ZK-BAH	John Baynes	John Baynes, Otama, Southland
de Havilland DH82A Tiger Moth	ZK-BLI(NZ1448)	Sir Peter Elworthy	Sir Peter Elworthy, South Canterbury
de Havilland DH82A Tiger Moth	ZK-BRB(NZ1459)	Andy Woods	Alpine Fighter Collection, Wanaka
de Havilland DH82A Tiger Moth	ZK-BRC	Jules Tapper	Mount Cook Group, Queenstown
de Havilland DH82A Tiger Moth	ZK-BUO(NZ795)	Simon Spencer-Bower	Simon Spencer-Bower, Claxby, Canterbury
de Havilland DH89A/B Dragon Rapide/Dominie 'Tui'	ZK-AKY	Neil Robertson	R. Gerald Grocott, Mandeville
de Havilland DH104 Devon NT.1	ZK-KTT(NZ1808)	Chris Bellamy, Peter Brown	Dalziell/Brown Syndicate, Ardmore
de Havilland DH104 Devon NT.1	ZK-RNG(NZ1807)	Glynn Powell	A.J. & R.M. Nicholson, Palmerston North
de Havilland DH104 Devon C.1	ZK-UDO(NZ1821)	Ray Robinson, Brian Horne	Devon Syndicate, Ardmore
de Havilland DH115 Vampire T35W	ZK-VAM(A79-649)	Ross Ewing, John Peterson	Vampire Syndicate, Ardmore
de Havilland Canada DHC-1 Chipmunk Mk 21	ZK-MUH(WB568)	Simon Spencer-Bower	Alpine Fighter Collection, Wanaka
de Havilland Canada DHC-1 Chipmunk Mk 22A	ZK-TNR(WB566)	Ian Reynolds	DH Chipmunk Syndicate, Ardmore
de Havilland Canada DHC-2 Beaver	ZK-CKH(NZ6001)	Peter Rhodes	NZ Warbirds Syndicate, Ardmore
Denney Aerocraft Kit Fox III	ZK-KNZ	Dan Denney	Messrs. Knauf, Taylor & Vowles, Wanaka
Douglas C-47/DC-3 Dakota	ZK-DAK(26840)	Bill Howell, Peter Creedon	Flightline Aviation, Ardmore
Fletcher FU24-950M	ZK-CBD	Mark Houston	Fieldair Holdings Ltd, Palmerston North
Glaser-Dirks DG400 (motorised glider)	ZK-GNU	David Speight	J.G. Speight, Te Anau
Grumman G-164A Ag-Cat	ZK-MEW	Phil Maguire	Pionair Adventures, Queenstown

Aircraft	Registration	Pilot(s)	Owner(s) and Base
Hawker 'Baghdad' Sea Fury Mk FB.11	ZK-SFR(WJ232)	Robbie Booth	Fury Syndicate, Ardmore
Isaacs Fury II (7/10 scale replica)	ZK-JHR(K2059)	Rex Carswell	Rex Carswell, Ardmore
Messerschmitt 109J (Hispano Aviacion HA-1112-M1L Buchon)	G-BOML	Mark Hanna	Old Flying Machine Company, Duxford, England
North American AT-6C Harvard Mk IIA	ZK-ENA(NZ1037)	Ross Ewing	John Mathewson, Ranfurly, Otago
North American AT-6D Harvard Mk III	ZK-ENC(NZ1091)	Ken Walker	W.J.D. & S.M.A. Williams, Tauranga
North American AT-6D Harvard Mk III	ZK-ENE(NZ1066)	Robbie Booth, Bill Rolfe	Booth/Porter/Rolfe Syndicate, Ardmore
North American AT-6D Harvard Mk III	ZK-ENF(NZ1065)	John Lamont, Bill West	Broadbent/Duncan Syndicate, Ardmore
North American AT-6D Harvard Mk III	ZK-ENG(NZ1078)	Steve Taylor, Ken Walker	Harvard 78 Syndicate, Auckland
North American AT-6D Harvard Mk III	ZK-ENJ(NZ1098)	John Denton	K.R. Brooking, Tauranga
North American AT-6D Harvard Mk III	ZK-ENK(NZ1099)	Keith Skilling, Jim Pavitt	Charles Darby & Jim Pavitt, Ardmore
North American AT-6C Harvard Mk IIA	ZK-SGQ(NZ1033)	John Lanham	NZ Sport and Vintage Aviation, Masterton
North American AT-6D Harvard Mk III	ZK-WAR(NZ1092)	John Peterson, Peter Houghton	Bland/Schroder Syndicate, Ardmore
New Zealand Aerospace Industries NZAI Cresco 08-600	ZK-LTQ	Malcolm Hill	Turbo Air Services Ltd, Alexandra
Pilatus PC6/B2-H4 Turbo Porter	ZK-MCN	Wayne McMillan	The Mount Cook Group, Queenstown
Piper PA-18A-150 Super Cub	ZK-BNY	Phil Murray	J.B & B.S. Murray, Tekapo
Pitts Special S1	ZK-EES	John Kelly	R.& M. Hood and J. Kelly, Ardmore
Replica Plans S.E.5a (7/10 scale replica)	ZK-TOM(B168)	Tom Grant	Tom Grant, Taieri
Vans RV-3	ZK-PMH	Phil Higgins	P.J. Higgins, Rotorua
Vickers Supermarine Spitfire Mk LF XVIe	ZK-XVI(TB863)	Ray Hanna, Tom Middleton	Alpine Fighter Collection, Wanaka

Parachutists

(in jump order) Peter Willemse, Brent Findlay, Brett Kennedy, Graeme Bates, Rob Nisbet, Avril Dolman, Alistair Kay, Mike Humphrey, Kate Bristow; of the Southland Parachute Club and Otago Skydiving Club

Interesting Statics

Aero Engine Service Ltd AESL CT-4 Airtrainer	ZK-DGY	CT-4 Syndicate, Ardmore (the prototype '01')
Auster J/5F Aiglet Trainer	ZK-BRA	D.K. Edmonds, Taieri
de Havilland DH82A Tiger Moth	ZK-BLV	Mrs Margaret Thelwell, Hastings
Mignet HM 14 Pou du Ciel (1933 Flying Flea) Engine: Scott Flying Squirrel	ZM-AAC	R. Gerald Rhodes, Wanaka (unassembled)
Lowther Nieuport 17-bis (7/8 scale replica)	ZK-NIE	John Lowther, Timaru
Pazmany PL-4A	ZK-PLF	J.D. Kramer, Dunedin
Rand Robinson KR-2	ZK-JAY	L. Day, Alexandra
Stodart-Hamilton SH-2 Glasair RG	ZK-NRG	Greg J. Laird, Auckland
Yeoman YA-1 Cropmaster 250R	ZK-CPW	Southern Aviation Ltd, Gore

Helicopters

Aerospatiale AS350B Ecureuil (Squirrel)	ZK-HMY	Bill Black	The Helicopter Line, Te Anau
Hiller UH-12E 'Herbie'	ZK-HBL	Tim Wallis	Alpine Deer Group, Wanaka
Hiller UH-12E	ZK-HDM	Davida Mead	T.G. & D.I. Mead, Dingleburn
Hughes 269C Model 300C	ZK-HGP	Peter Paterson	H.E. Clementson Ltd, Westport
Hughes 369HS Model 500C	ZK-HQE	Kim Hollows	Kim Hollows, Te Anau
Hughes 369D Model 500D	ZK-HOT	Tim Wallis	Alpine Deer Group, Wanaka
Hughes 369D Model 500D	ZK-HWH	Dennis Egerton	The Helicopter Line, Queenstown
Hughes 369E Model 500E	ZK-HFT	Brian Hore	B.L. & M.A. Hore, Lumsden
Robinson R22 B Beta	ZK-HDY	Doug Maxwell	D.M. Maxwell, Alexandra
Robinson R22 B Beta	ZK-HHH	Harvey Hutton	Harvey A. Hutton, Wanaka

Royal New Zealand Air Force

Aermacchi MB-339CB 'Macchi'
No. 14 Squadron, RNZAF Base Ohakea

	NZ6468	Sqdn.Ldr. Bruce Keightley (Herb) (static only)

Aero Engine Service Ltd AESL CT-4B Airtrainers
Central Flying School and Pilot Training Squadron, RNZAF Base Wigram

Royal New Zealand Air Force — continued

'Red Checkers'	Red 1	NZ1938	Sqn.Ldr. Ian McClelland 'Mac'
	Red 2	NZ1935	Sqn.Ldr. Hoani Hipango 'Hippo'
	Red 3	NZ1936	Flt.Lt. Nick Osborne 'Oz'
	Red 4	NZ1942	Flt.Lt. Phil Murray 'Phil'
	Red 5	NZ1931	Flt.Lt. Mike Tomoana 'Tux'
	Red 6	NZ1945	Flt.Lt. Peter Walters 'Pee Dub'
			Reserve/Commentator

Bell UH-1H Iroquois
No. 3 Squadron Detachment, RNZAF Base Wigram

| | NZ3805 | Flt. Lt. Mark D. Cook, Flg.Off. Anthony F. Fesche, F/SHCM Terry L. Houghton |

British Aircraft Corporation BAC 167 Strikemaster Mk 88
No. 14 Squadron, RNZAF Base Ohakea

| | NZ6361 | Flg.Off. Shaun Singleton-Turner (static) |
| | NZ6363 | Flt.Lt. Mike Longstaff (display pilot) RAF exchange pilot with 14 Squadron |

Fokker F27-100 Friendship
Navigation, Air Electronics and Telecommunications Training Squadron, RNZAF Base Wigram

| | NZ2782 | Flt.Lt. Jon Wangford ('Babs'), Flg.Off. John Harding ('J.H.') (static) |

Hawker-Siddeley Andover C Mk 1
No. 42 Squadron, RNZAF Base Auckland

| | NZ7620 | Flt.Lt. Neil Kenny, Flg.Off. Derek Singer, Flg.Off. Dane Fea, S/ALM. Brett Leong |

Kiwi Blue Parachute Team

Captain John Tinsley, Flt.Lt. Severn Smith, Flt.Lt. Paul Neave, Sgt. Grant Murdoch,
Sgt. Keith Dickinson, Sgt. Colin Stewart, Sqn. Ldr. Bob Howard (drop zone controller on ground)

Appendix 5
Fly-in 1993

Aircraft	Registration	Pilot(s)	Owner(s) and Base
Altavia PZL 104 Wilga 35A	ZK-PZO	Dougal Dallison	Rick Maclean, Wanaka
Chance Vought F4U-1 Corsair	ZK-FUI(NZ5201)	John Lamont, Tom Middleton, Keith Skilling	Alpine Fighter Collection, Wanaka
Curtiss P-40K-5-CU Kittyhawk	ZK-PXL(NZ3108)	Static	Alpine Fighter Collection, Wanaka
de Havilland DH82A Tiger Moth	ZK-BAH	John Baynes	John Baynes, Otama, Southland
de Havilland DH82A Tiger Moth	ZK-BRB(NZ1459)	Static	Alpine Fighter Collection, Wanaka
de Havilland DH82A Tiger Moth	ZK-BCO	Static	George Kingsbury, Wanaka
de Havilland Canada DHC-1 Chipmunk	ZK-MUH(WB568)	Tim Wallis	Alpine Fighter Collection, Wanaka
de Havilland DH115 Vampire Mk T35W	ZK-VAM(A79-649)	Brett Emeny	Vampire Syndicate, Ardmore
Grumman TBM-3E Avenger	ZK-TBM(NZ2518)	Rex Dovey, Phil Murray, Tom Middleton	Alpine Fighter Collection, Wanaka
Canadian Car & Foundry Company Harvard IV/A6M2 'Zeke'	N15798(288)	Phil Murray, Simon Spencer-Bower	Alpine Fighter Collection, Wanaka
North American AT-6C Harvard Mk IIA	ZK-ENA(NZ1037)	John Mathewson	John Mathewson, Ranfurly, Otago
North American P-51D Mustang	ZK-TAF(NZ2415)	Trevor Bland	John Sager and Graham Bethell, Ardmore
Pitts Special S-2A	ZK-PTO	Grant Bisset	Biplane Adventures, Wanaka
Hughes 369D Model 500D	ZK-HOT	Tim Wallis	Alpine Deer Group, Wanaka

Bibliography

Newspapers:
Southland Times
Otago Daily Times

Periodicals:
New Zealand Wings, 1981 to present, especially New Zealand Warbirds Directory, February 1992–May 1992.

Booklets:
The Royal New Zealand Air Force — Yesterday and Today, edited by RNZAF Public Relations, Wellington, Flight Lieutenant Diana Hales and Sergeant Rob Salmon, GP Ltd, 1985.

Books:
Constable, Trevor J. and Toliver, Colonel R.F. (ret.), *Horrido!: Fighter Aces of the Luftwaffe*. Arthur Baker Ltd, 1968.
King, John, *Vintage Aeroplanes in New Zealand*. Heinemann Publishers, 1986.
Ladd, Captain Fred, MBE and Annabell, Ross, *A Shower of Spray and We're Away*. A.H. & A.W. Reed Ltd, 1971.
Rawlings, John D.R., *Fighter Squadrons of the RAF and Their Aircraft*. MacDonald & Janes, 1976.
Roxburgh, Irvine, *Wanaka and Surrounding Districts: A Sequel to the Wanaka Story*. Publication Committee of Upper Clutha 1990 Community Committee, 1990.
Williamson, Gordon, *Knights of the Iron Cross: A History 1939–1945*. Blandford Press, 1987.

The souvenir programmes from the 1988, 1990 and 1992 airshows were also an invaluable source of information.

Photographic Credits

The author gratefully acknowledges the assistance of the 19 photographers who have provided photographs for this book. They are listed below with the page numbers on which their photographs appear. The remaining photographs were taken by the author.

Barry Harcourt, *Southland Times*, Invercargill   back cover, pp. 18/19, 48 bottom right, 50/51, 55 bottom, 60 bottom, 63, 76, 79, 88, 91, 108 left, 111, 113 top left, 123.

Flt. Sgt. Michael Provost, RNZAF Wigram   pp. 62 top and bottom, 65, 83 bottom, 89, 99, 103, 105, 107, 112.

David Wethey, *The Press*, Christchurch   pp. 10, 20, 23, 26, 28, 30/31, 68 bottom, 78, 80 .

Gerard O'Brien, *Otago Daily Times*, Dunedin   pp. 57, 74, 75 bottom, 86, 92/93, 100 top and bottom, 104, 124.

Alistair Kinniburgh Photo Ltd, Christchurch   front cover, pp. 2, 64 top and bottom, 71 top, 97.

Ross Macpherson, *New Zealand Wings*   pp. 21, 40/41, 44 top, 46 bottom, 73, 95 bottom.

Alan Gibb, Invercargill   pp. 24/25, 29 top, 60 top, 68 top, 101.

Graeme Thorburn, Dunedin   pp. 29 bottom, 32/33, 72, 114/115.

Prue Wallis   pp. 53, 117.

Scenix Photo Library   pp. 14/15.

Ian Brodie, New Zealand Fight Pilots Museum, Wanaka   p. 46 top.

David Richards, Christchurch   p. 47.

Cpl. Darryl Thorburn, RNZAF Base Wigram   p. 90.

Marcus Thorburn, Dunedin   p. 96.

John McDonald, Christchurch   p. 102.

Mark Hampton, Hokitika   p. 106.

Stefan Thorburn, Dunedin   p. 108 right.

Flt. Sgt. Paul Hillier, RNZAF, Wellington   p. 110.

Lyndon McEntee, Christchurch   p. 118.